The NAG Library:
A Beginner's Guide

The NAG Library: A Beginner's Guide

JEN PHILLIPS

Senior Lecturer in Mathematics
Middlesex Polytechnic

CLARENDON PRESS · OXFORD
1986

Oxford University Press, Walton Street, Oxford OX2 6DP
Oxford New York Toronto
Delhi Bombay Calcutta Madras Karachi
Petaling Jaya Singapore Hong Kong Tokyo
Nairobi Dar es Salaam Cape Town
Melbourne Auckland
and associated companies in
Beirut Berlin Ibadan Nicosia

Oxford is a trade mark of Oxford University Press

Published in the United States
by Oxford University Press, New York

British Library Cataloguing in Publication Data
Phillips, Jen
The NAG Library: a beginners guide.
1. NAG (Computer programs)
I. Title
005.3   QA76.6
ISBN 0–19–853263–6

Library of Congress Cataloging in Publication Data
Phillips, Jen, 1938–
The NAG Library.
Includes index.
1. NAG Library (Computer programs)   2. Numerical
analysis—Computer programs.   3. FORTRAN (Computer
program language)   4. Subroutines (Computer programs)
I. Title.
QA297.P53   1986   515'.028'553   86-8746
ISBN 0–19–853263–6

Set by Wyvern Typesetting Ltd, Bristol
Printed in Great Britain by
St Edmundsbury Press Ltd, Bury St Edmunds, Suffolk

Preface

The main purpose of this book is to introduce the reader to the Fortran version of the *NAG* (Numerical Algorithms Group) *Library*.

The *NAG Library* consists of a large collection of subroutines and functions. Each routine does a specific numerical job, such as solving differential equations, numerical integration, solving linear programming problems, and so on.

A detailed description of how to use each NAG subroutine is provided in the *NAG Library Manual*. These descriptions are complete, precise and logical, but can be daunting or confusing to a newcomer to the Library. This book aims in the first instance to give an alternative 'eased-up' description of a selection of these routines. It also aims to give the reader enough information and experience to be able to use the *NAG Library* with confidence.

The book is split into three parts. Part 1 consists of miscellaneous introductory material, which should be read by anyone wishing to use the NAG routines. Part 2 consists of a selection of eased-up descriptions of NAG routines, including some NAG graphical routines. Part 3 consists of an introduction to the use of the *NAG Manual*.

Many people have been dragged willingly and less willingly into the compilation of this book. Special thanks are due to the following people and Institutions:

NAG (Numerical Algorithms Group), without whose fullest co-operation this book could not have been written;

Middlesex Polytechnic, which has provided me with all the test facilities I have needed;

Jeremy du Croz (NAG) and *John Phillips* (Middlesex Polytechnic) for the great amount of help and support they have provided;

R. A. Bailey (Rothamsted Experimental Station), *Peter Dean* (Institute of Education, University of London), *Oliver Penrose* (Open University), *Brian Smith* (Argonne National Laboratory), and others, whose comments on the draft of this book have helped me to put the final form of the book together. Last, but not least, many thanks to the numerous students at Middlesex Polytechnic who have tested and commented on the material in this book.

Middlesex Polytechnic   J. P.

v

To obtain machine-readable copies of the specimen programs printed in this book, please write to:

> The Numerical Algorithms Group Ltd
> Mayfield House
> 256 Banbury Road
> Oxford
> OX2 7DE
> England

or, in North America, to:

> The Numerical Algorithms Group Ltd
> 1101 31st Street, Suite 100
> Downers Grove
> Illinois 60515–1263
> USA

Contents

Part 1
Introduction

This part of the book includes miscellaneous information needed to be able to successfully run the NAG routines described both here and in the *NAG Manual*.

Included here is

- general information about the *NAG Library*

- a description of particular aspects of Fortran which are needed to use the NAG routines

- programming conventions used in the NAG routines

- three specimen NAG routines, along with specimen programs which call these routines

- notes on how to avoid mistakes, and so on.

To make maximum use of this book, read Part 1 carefully before proceeding further. It is recommended that you run the three specimen programs. This will give you the opportunity to find out the changes which will have to be made to subsequent programs to ensure that they work on your particular computer.

1
Some questions (and answers)

Q Who is this book for?

A For scientists, engineers, students – in fact, for anyone who needs a gentle introduction to the numerical, graphical, or statistical routines in the *NAG Library*.

Q What is NAG?

A 'NAG' stands for 'Numerical Algorithms Group'. NAG began in 1970 when a group of numerical analysts in British universities began to work together to develop a library of subroutines (referred to as the *NAG Library*) for numerical computation. The project was successful and continued to grow. Other institutions became interested in using the *NAG Library*, and more and more people became involved in contributing to the Library and in testing it on different types of computers. To provide a proper organization for this activity, a non-profit-making company, NAG Ltd., was established in 1976, based in Oxford. Later a subsidiary company, NAG Inc., was established in the USA. However, although the staff of these companies play a very important role, they are not the whole of NAG. Many experts, in universities and research laboratories all over the world, now work with NAG by contributing to the *NAG Library* and other software distributed by NAG.

Q How wide is the scope of the *NAG Library*?

A The Library contains routines to do most of the standard numerical jobs (such as solving differential equations, finding eigenvalues, and integration). It also contains some statistical routines, and a set of graphical routines in a separate supplement. This book contains descriptions of a selection of the routines available.

Q What language is used to call these routines?

A By far the most heavily used version of the *NAG Library* is the Fortran version. The routines in it are intended to be called from programs written in Fortran. At some sites it is possible to call NAG Fortran routines from programs written in other languages (e.g. Pascal), and there are also Algol 68, Algol 60, and Pascal versions of some of the routines. This book, however, concentrates on the Fortran version.

Q How much mathematics is needed to be able to use the NAG routines?

A Not a great deal usually. You need, however, to be able to state the problem which you want to solve in suitable mathematical terms, and

3

occasionally you need some understanding of the mathematical or numerical principles, in order to run the routine in a sensible way. Where this is the case for routines described in this book, then the necessary information is included in the routine description. However, if you can't get the routine to work properly, or the results are unsatisfactory, you must be prepared to get some expert help.

Q How much Fortran is needed to be able to use the NAG routines?

A Again, not a great deal. It is assumed for the purposes of this book that you know enough Fortran to be able to read in data and to print out results (including one- and two dimensional arrays), and that you can write a program with a loop in it. If you are unable to do this, then you ought first to look at an introductory book on Fortran. However, there are a few aspects of Fortran programming which arise frequently when using NAG routines. These are described briefly in the Fortran notes in Chapter 2.

Q Is the *NAG Library* widely used?

A Yes. It is the predominant library of numerical subroutines in the United Kingdom, and is used in universities, polytechnics, government laboratories, and industrial companies. It is also widely used in western Europe and North America. Altogether it is used in about 1000 institutions in over 40 countries around the world. So, once having learnt to use the *NAG Library*, there is a good chance that you will be able to continue to use it in your subsequent career.

Q Can the *NAG Library* be used on any type of computer, including microcomputers?

A Almost any. NAG has taken a lot of trouble to make the Library 'transportable': that means, making it easy to implement on different types of computers. NAG also makes a point of testing the Library thoroughly on each type of computer before making it available. Currently, the *NAG Library* is available on over 30 different types of computer, and the number is steadily growing.

The *NAG Library* as a whole is simply too big to be used on most microcomputers. However, NAG have selected 50 of the most commonly used routines, and made them available as a separate small library (called the PC50) on some of the smaller microcomputers. For more powerful microcomputers and scientific workstations, NAG is planning to distribute a larger selection of about 150 routines. About half the routines described in this book are included in the PC50, and almost all of them will be included in the larger selection.

Q How do I use the NAG routines?

A That is what the book is about . . .

2
Fortran notes

The following notes describe various aspects of Fortran. These aspects are needed and used in the implementation of most NAG routines, both in this book and in the *NAG Manual*.

2.1 Fortran subroutines and functions

Fortran allows for two kinds of subprograms – *subroutines* and *function subprograms* (or just 'functions'). Both functions and subroutines are often loosely referred to as 'routines'. The *NAG Library* consists of a collection of routines, most of which are subroutines. However, there are also a few functions (such as the Bessel function routines described in §6.15). So, before you can use the *NAG Library*, you have to know how to 'call' a subroutine, and how to use a function. Some NAG routines also require you to write a subroutine or a function yourself, which is to be used by the NAG routine. So you also need to know how to construct both types of routine.

It is for these reasons that a brief account of how to construct and use subroutines and functions is included in this book. If you are familiar with both types of routine, then skip the rest of this section.

(a) *Subroutines*

A subroutine is a self-contained set of instructions – a subprogram which can be called, many times if necessary, by another program. For instance, suppose that you wanted to put the instructions to add and subtract two numbers, A and B, into the form of a subroutine. The instructions themselves can be written

$$R = A + B$$
$$S = A - B$$

A subroutine containing these instructions could be written as follows:

```
SUBROUTINE ADSUB (A, B, R, S)
REAL A, B, R, S
R = A + B
S = A - B
RETURN
END
```

Note that every subroutine has a name. The name of the subroutine above is ADSUB. Also, all subroutines should have an instruction

RETURN

in them.

Although subroutines have most of the properties expected of a program, they cannot be run on their own. They can be used only by 'calling' them from another program.

Your program   A subroutine

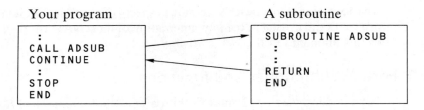

If you write an instruction

CALL ADSUB

in your program, then the computer jumps to the subroutine ADSUB and follows the instructions in it until it reaches the instruction.

RETURN .

At that point, control is then passed back to your program (the '*calling program*'), and the computer continues by executing the instruction immediately following the CALL statement. There may be many such CALL instructions in your program: the computer 'remembers' which one it last encountered, and returns control to the right place.

This is the basic story, but it is not quite enough.

Parameters (or arguments) of a subroutine

In the case of the subroutine ADSUB above.

(a) the subroutine needs the values of A and B from the calling program so that it can perform its operations, and

(b) the calling program needs the 'answers' R and S back from the subroutine.

To achieve this, both the first line of the subroutine and the call statement in the calling program can be extended to include the names of the variables whose values have to be communicated.

(a) The first line of the subroutine could be written

SUBROUTINE ADSUB (A, B, R, S).

The list of variables in parentheses are either those values (i.e. A and B) which are required by the subroutine or those values (i.e. R and S) which are required afterwards by the calling program. These variables A, B, R, and S are referred to as *arguments* or *parameters*, and the list (A, B, R, S) is called the *argument* or *parameter list*. In keeping with current NAG practice, these variables will be referred to as parameters in this book.

To be able to call this subroutine (or any other), *you have to know the meaning of the parameters, and the order in which they occur in the parameter list*.

(b) Suppose now, that you wanted to use the subroutine ADSUB to find C + D and C − D, putting the answers in X and Y respectively. To do this, a statement

CALL ADSUB (C, D, X, Y)

should be used in the calling program. Then, when the program is run, and the call statement above is met, each variable in the list (C, D, X, Y) above is matched with the parameter list (A, B, R, S) of ADSUB, according to the order in which they appear. So C is matched with A, D with B, and so on, and the computer effectively executes the instructions X = C + D and Y = C − D.

Thus, the parameter list which follows a subroutine name provides a means by which information can be communicated between a calling program and a subroutine.

A more detailed description of the use of subroutines can be found in any good book on Fortran.

(b) *Functions*

The purpose of a Fortran function subprogram (or routine) is to provide a convenient way of packaging a set of instructions which compute a *single* result. An obvious use of a function routine is to compute the value of a mathematical function.

Suppose, for instance, that you needed to use a function $f(x) = \sin(x^2)$ in your program. Then you could write the following function routine at the end of your program:

```
REAL FUNCTION F(X)
REAL X
F = SIN(X*X)
RETURN
END
```

Note that functions, like subroutines, have a name, parameters, and a RETURN statement. In the routine above, the function routine has a name, F, and in this case, just one parameter, X.

However, unlike a subroutine, it is the value of the function (rather than the parameter values) which is passed back to the main program. Thus, in the example above, it is the value assigned to F which is returned to the main program. A function routine with a name, F say, must have a statement 'F = . . .' before the RETURN statement.

Suppose that you wanted to use the function F above to evaluate SIN(A^2). Then you would simply write a statement

   Y = F(A)

in your program.

When the program is run, the function F is found, and evaluated for X = A in the statement

   F = SIN(X*X).

Finally, the value SIN(A*A) is returned to Y.

A more detailed description of function subprograms can be found in any good book on Fortran.

2.2 Supplying a routine as a parameter

Subroutines in general (and NAG routines in particular) may need more than values of variables from a calling program. They might also require information about functions, or even other subroutines.

For instance, a subroutine SPEC which evaluates the integral

$$\text{ANS} = \int_a^b f(x) \, \mathrm{d}x$$

not only needs the values a and b from a calling program, but also requires the user to provide the function f which is to be integrated. Assuming that the routine writer wants to pass the answer back to the calling program, then a suitable first line for the subroutine SPEC might be

   SUBROUTINE SPEC (A, B, F, ANS).

So, as you can see, function names may be included in parameter lists, where the user is left to provide the function. To do this

   (a) your calling program must have a declaration

      EXTERNAL F

   (b) you must provide a function routine F(X).

So, if you wanted to use the subroutine SPEC to evaluate the integral $\int_a^b \sin(x^2)\mathrm{d}x$, then a suitable program could be written as follows:

```
REAL A, B, F, ANS
EXTERNAL F
READ (5,*) A, B
CALL SPEC (A, B, F, ANS)
WRITE (6,*) ANS
STOP
END

.REAL FUNCTION F(X)
REAL X
F = SIN (X*X)
RETURN
END
```

Comments:

(i) The EXTERNAL statement above indicates that F is the name of a routine rather than the name of a variable; it could, in fact, be either a function or a subroutine.

(ii) The function can be given any name you choose, but make sure that the same name is used in the EXTERNAL statement, in the CALL statement, and in the function itself.

(iii) There is nothing special about using X as a parameter. Again, any name (subject to Fortran naming conventions) would do.

2.3 Declaration of variables

It is the policy, both of this book, and of the *NAG Manual* to start the specimen programs with a declaration of all the variables, functions, and subroutines which are going to be used in that program. Thus, a typical specimen program might well start with statements like

```
REAL P, A(10,10), L
INTEGER I, R, D(2)
COMPLEX C
LOGICAL STAGE
EXTERNAL F
```

Comments on the statements above

(i) The statements

REAL L
and INTEGER R

ensure that, for the purposes of the program in which these statements appear, L is real and R is integer.

(ii) The statements

REAL A(10,10)
and INTEGER D(2)

ensure that a real 10 × 10 two-dimensional array A, and an integer one-dimensional array D of length 2 can be used in the program.

(iii) The statements

COMPLEX C
and LOGICAL STAGE

ensure that C can be used to contain a complex number, and that STAGE is a logical variable.

(iv) The statement

EXTERNAL F

ensures that F is the name of the function or a subroutine.

Now, the statements in the comments above are all essential, whereas the statements

REAL P
and INTEGER I

are not. The standard rule for variable names in Fortran is that, *unless declared otherwise*, any variable name beginning with I, J, K, L, M, and N will be an integer, and all others will be real. This rule can be overridden by using statements like the ones above. There are two obvious advantages of declaring all the variables, irrespective of whether it is necessary or not:

(i) The habit minimizes the chance of omitting one of the essential declarations.

(ii) It is essential in some other programming languages (notably, Pascal and Algol) to declare all the variables to be used in a program. So, the inclusion of these declarations in a Fortran program facilitates the translation of a program from one language to another, if necessary.

2.4 Dimensions of arrays in subroutines

In a Fortran program, the size of arrays in subroutine declarations must be less than or equal to the size of the corresponding arrays in the declaration in the calling program.

Thus, if a calling program has a declaration

REAL A(4), B(5) ,

then the Fortran language insists that the equivalent declaration in the subroutine has the form

REAL A(a), B(b)

where a and b must be positive integers not greater than 4 or 5 respectively.

To enable this restriction to be satisfied, Fortran allows the use of 'adjustable' dimensions in subroutines.

Suppose, for example, a subroutine has a name

SPEC2 (M, N, . . .) .

Then Fortran allows SPEC2 to have a declaration

REAL A(M), B(N) .

Thus, in the example above, if M and N are parameters of SPEC, where suitable values for M and N are provided in the calling program, then the lengths of the arrays A and B in the subroutine will both be suitable for your particular problem, and will also satisfy the Fortran restriction.

Furthermore, in the case of a two-dimensional array, the declaration of the first dimension size must be the *same* in both the calling program and the subroutine.

So, if the calling program has a declaration

REAL A(10,8)

then the corresponding declaration in the subroutine must be of the form

REAL A(10,c)

where $c \leqslant 8$.

This declaration can take the form

REAL A(IA,c)

so long as IA is a parameter of the subroutine and IA has already been given the value 10 before the routine is called.

It is for this reason that you are often asked to supply the first dimension size of a two-dimensional array as a parameter in a NAG subroutine.

Suppose, for example, that A, B, and C were three arrays which were parameters of a NAG subroutine, and that you had a declaration

REAL A(10, 8), B(5, 7), C(8, 5)

in your calling program. Then in most NAG routines of this kind you would be asked to supply the value of three further parameters IA, IB, and IC (say). These must be the size of the first dimension of the arrays A, B and C respectively. So, before you called the routine, you would in this case explicitly give the values

IA = 10
IB = 5
IC = 8.

2.5 COMMON statements

In Fortran, there are two ways in which values can be passed from a calling program to a subroutine (or vice versa). One way is to use a parameter list, which is described in §2.1. The other way is to make a statement, that you wish for a variable (say A), in the calling program to have the same value as a variable (say C), in a subroutine. This is done by using COMMON statements

        COMMON A

in the calling program, and

        COMMON C

in the subroutine.

Thus, using the example given in §2.1, the two programs

```
REAL A,B,R,S
READ(5,*)A,B
CALL ADSUB(A,B,R,S)
WRITE(6,*)R,S
STOP
END

SUBROUTINE ADSUB(C,D,X,Y)
REAL C,D,X,Y
X=C+D
Y=C-D
RETURN
END
```

and

```
REAL A,B,R,S
COMMON A,B,R,S
READ(5,*)A,B
CALL ADSUB
WRITE(6,*)R,S
STOP
END
SUBROUTINE ADSUB
REAL C,D,X,Y
COMMON C,D,X,Y
X=C+D
Y=C-D
RETURN
END
```

would produce, from the point of view of the user, exactly the same effect. It goes practically without saying that the order in which variables are declared in common is all-important. In the example give above, A and C will have common values, so will B and D, and so on.

Use of COMMON statements is essential when you need to pass values, which are not parameters, to a subroutine. A situation like this can arise if you are asked to provide a subroutine, which is subsequently to be used by a NAG routine. In this case, the NAG routine writer will have specified what parameters this subroutine should have. However, you might need to introduce some extra information from the calling program in order to be able to write a satisfactory subroutine. In this event, you would have to put COMMON statements in the calling program and the subroutine. This book gives one example of essential use of COMMON statements in §10.7, and another example in the addendum to §6.4.

3
Introduction to the NAG routine descriptions

3.1 Layout of the descriptions

The descriptions of the NAG routines which are given in this book will be treated under the following headings:

(i) the **purpose** of the routine;
(ii) a **specimen problem**;
(iii) the **method** used;
(iv) the **routine name**;
(v) a **description of parameters**;
(vi) a **specimen program** and
(vii) a **postscript**.

(i) Under the heading '**purpose**', you will be told what the routine is doing. For instance, in the routine discussed in the next chapter, you are told that the purpose of the routine C02AEF is to find all the solutions of the polynomial equation

$$a_1 z^{n-1} + a_2 z^{n-2} + \ldots + a_n = 0 \ .$$

(ii) A **specimen problem** is given. This is used for demonstration purposes in the routine description, and is also the problem which is solved, using the specimen program.

(iii) Under the heading '**method**', you will briefly be told what method the routine uses. It is assumed that if you want or need more details, you will consult the *NAG Manual* for references.

(iv) Under the heading '**routine name**', the name of the routine will be given, followed by a list of parameters required for this routine. A typical example of this initial description is the one given for C02AEF in the next chapter. In that case the **routine name** with parameters is

C02AEF (A, N, ZRE, ZIM, TOL, IFAIL).

(v) The **description** section gives details of each parameter used in the routine. It will also describe any other functions or subroutines which the particular NAG routine may require.

Each parameter will be described in one of the following sections:

13

(a) Parameters which require values before the routine is called.

(b) Functions or subroutines which require definition.

[In this section, all the functions or subroutines needed by the NAG routine will be described, whether they are parameters or not.]

(c) Parameters associated with workspace.

(d) The error parameter.

(e) Parameters to be examined after calling the routine.

(f) Any other parameters.

Some, but probably not all, of these sections will occur in the description of each NAG routine.

(vi) Under the heading '**Specimen program**', you will find a program plan followed by a program. In the 'specimen run', the program is used to solve the specimen problem given at the beginning of the description of the routine.

(vii) Under the heading '**Postscript**', you will find (where appropriate)

- (a) comments on how the routine can be adapted to related problems.

(b) comments on the reliability of the answer obtained, and

(c) pointers to other related NAG routines.

3.2 The error parameter

A parameter which occurs in nearly all the NAG routines is one which indicates whether any trouble was found by the routine when it tried to run your problem. This parameter is called IFAIL in most of the routines, and works as follows:

If a NAG routine has found no trouble when running your problem, the routine sets IFAIL = 0. However, if the NAG routine detects an error, then IFAIL is assigned some non-zero value. In a simple routine, there may be only one basic error, in which case, if the error were detected, then the routine would set IFAIL to 1. The different values that IFAIL can take and the corresponding errors are listed in the description of each routine.

Before a NAG routine is called, IFAIL must be assigned a value. *You are recommended at this stage to set IFAIL = 0 before you call a NAG routine*. If you set IFAIL = 0 initially, then should anything go wrong in the routine, the routine will not only set IFAIL to a non-zero value, but also print it. Your program will then stop. At the end of §4.2 there is an example of an error message printed by a NAG routine.

An alternative way of using IFAIL is given in §5.3, but most of the specimen programs in this book use IFAIL in the way described above.

3.3 Workspace

In some NAG routine descriptions, there are parameters (usually one- and two-dimensional arrays), which are *'used as workspace'*. This means simply that the NAG routine itself needs these arrays to work in. In this case, all you have to do when writing your program is to *declare these arrays with the dimensions specified in the routine description, and to pass their names* (and sometimes their dimensions) *as parameters in a call to the routine*. You do not need to assign any values to them, and usually you do not need to examine their contents after the routine has been called.

3.4 Notes on the use of the specimen programs

(i) In the description of each routine in this book, there is a specimen program and specimen run attached. These programs are all written in standard Fortran 77 for implementation on a particular machine, and are intended to be run at a terminal. However, they may not work on your computer exactly as they stand, though the changes needed to adapt them will be very small. There are two obvious changes which might have to be made to the printed specimen programs:

(a) The unit number (5) in the READ statement and (6) in the WRITE statement may need to be changed for your computer. So you should check at your computer centre (if you don't already know) which unit numbers indicate a terminal for READ and WRITE statements.

(b) When you first want to run a NAG routine, you should first check which version of the *NAG Library* is available on your computer. If it is a *double precision* version, then you will have to change the word 'REAL' to 'DOUBLE PRECISION' in all the declarations in the printed specimen programs. For the normal single precision versions of the Library, this doesn't have to be done.

If in any doubt about what changes to make, consult your computer centre.

(ii) One of the advantages of Fortran 77 is that it allows the user to omit FORMAT statements. In an effort to keep the specimen programs simple, this book avoids using FORMAT statements as far as possible. This does mean, however, that usually no control has been exercised over the number of significant figures given in the output. So care must be taken to interpret the answer intelligently. The postscript to the routine may advise you how many significant figures are likely to be meaningful in the context of the given program and specimen data.

(iii) Before you attempt to run a program which calls a NAG routine, *you will have to find the instruction(s) which are necessary to access the*

NAG routines from your program on your particular computer. This can be quite different for each machine, so go and check what to do at your computer centre.

(iv) All the programs in this book are written on the assumption that you are working 'on-line'. However, they can all easily be converted to 'batch-mode', or other use. This is discussed further in §4.1.

(v) When you run the specimen programs on your computer, you might observe that the answers you get differ slightly from the ones given in this book. This can be caused by the respective computers working to a different number of significant figures, or just a different number of significant figures being printed in the answer. *So slight differences between your answer and the printed answer are usually nothing to worry about.*

(vi) The programs in this book demonstrate a simple way of using the routines. It is expected that you will adapt these programs to your own use.

4
Specimen routine descriptions

In this chapter, three specimen routines are described. Due to their introductory nature, rather more comment is included here than in subsequent descriptions in this book.

It is recommended that you read all three descriptions, and run the specimen programs. These descriptions have been chosen to cover a number of features of Fortran which are commonly used in connection with the *NAG Library*.

4.1 Solutions of a polynomial equation: C02AEF

The **purpose** of this routine is to find all the complex solutions of the polynomial equation

$$a_1 z^{n-1} + a_2 z^{n-2} + \ldots + a_n = 0 \qquad (a_1 \neq 0) \qquad (4.1)$$

where the coefficients $a_1, a_2, \ldots a_n$ are real.

Comment:
Note that the equation above has n coefficients, and has degree $(n - 1)$. So, the routine finds $(n - 1)$ complex solutions of the equation. In fact, some of these solutions may be real.

Specimen problem

To find all the complex solutions of the equation

$$3.1z^5 + 2.2z^4 - 7.3z^3 - 1.4z^2 - 5.8z - 1.1 = 0 .$$

The **method** used is due to Grant and Hitchins.

Comment:
In the following routine descriptions, very limited information is given about the method. If more detail is required consult the *NAG Manual* for references.

The **routine name** with parameters is

C02AEF (A, N, ZRE, ZIM, TOL, IFAIL) .

Comment:

In general, routine writers do try to give meaningful names to their parameters. For instance, in the parameter list above, A stands for the coefficients $a_1, a_2, \ldots, a_n$ in equation (4.1), N for the n in (4.1), ZRE and ZIM for the real and imaginary parts of the solutions, and IFAIL for the error parameter described in §3.2. In no sense does this kind of guesswork let you out of reading the descriptions of the parameters carefully, but it does help.

Description of parameters

Parameters which require values before CO2AEF is called

N: [an integer variable]
N is the number of coefficients in the equation (4.1) above.

For instance, in the specimen problem, there are six coefficients, so you would have to specify the value N as 6.

Comments:

(i) N is restricted to having a value between 2 and 100.

(ii) Warning: N has its value changed by the routine. As you are going to need the original value of N to get the $N - 1$ solutions printed at the end, *you must make a copy of N (or N − 1) before you call the routine.* Note that the specimen program sets NSOL (the number of solutions) to $N - 1$, before the routine is called.

(iii) It is, in fact, quite unusual in NAG routines for parameters as important as N to have their value changed. So it is a moral lesson to read the parameter descriptions carefully.

A: [a real one-dimensional array. Its length must be *at least* N in the declaration in your calling program]

The array A is used to contain the coefficients $a_1, a_2, \ldots, a_n$ of the equation (4.1). In general, A(I) must contain the coefficient of z^{N-1}.

So, in the case of the specimen problem, where N = 6, A(1) should contain the coefficient of z^5, A(2) the coefficient of z^4, and so on. In this case you would specify the values

$$A(1) = 3.1 \quad A(2) = 2.2 \quad A(3) = -7.3 \quad A(4) = -1.4$$
$$A(5) = -5.8 \quad A(6) = -1.1 \,.$$

Comments:

(i) All arrays needed in a program *have* to be declared at the beginning of the program.

(ii) The reference '*at least N*' above is best explained by means of an example.

In the specimen problem, $N = 6$. In this case, you must have a *minimum* declaration

REAL A(6)

in your calling program.

This does not prevent you from making a larger declaration. For instance, a declaration

REAL A(10)

would allow you not only to solve this equation, but also to solve some higher degree equations without having to change the declaration in the program.

TOL: [a real variable]

TOL controls the accuracy of the solution.

You are advised to set

TOL = 0.0.

In this way, the solutions will be calculated to maximum accuracy.

[For further details, see the *NAG Manual*.]

The error parameter

IFAIL: [an integer variable]

IFAIL is the error parameter described in §3.2. It is recommended that you set

IFAIL = 0

before you call C02AEF. Then in the event of the routine failing, your program will stop and print one of the following messages:

Error message	Meaning	Advice
IFAIL = 1	*Either* $A(1) = 0$ *or* $N < 2$ *or* $N > 100$.	Check values of A(1) and N. [Note restriction on N.]
IFAIL = 2	Possibly the routine has encountered a point of inflection.	Get some help.

If you have any doubt as to what to do after the routine fails, get some expert help.

Parameters to be examined after calling C02AEF

ZRE⎫ [real one-dimensional arrays. Both arrays must have length at least N
ZIM⎭   in the REAL declaration in your calling program]

If all goes well in the routine, then the real and imaginary parts of the solution z_k of equation (4.1) will be found in ZRE(K) and ZIM(K) respectively.

Specimen program

Comment:
For each routine the specimen program solves the specimen problem stated at the beginning of the description.

Program planning

1. *Declare* REAL A( ), ZRE( ), ZIM( ), TOL
   INTEGER N, IFAIL

2. *Read* N, A
   Set   NSOL (the number of solutions), IFAIL, TOL

3. *Call* C02AEF

4. *Print* ZRE, ZIM

Comment:
(i) The following program will find solutions of polynomial equations of degree less than 10. If you wish to use this program for a polynomial of higher degree, then you will have to alter the REAL declaration at the beginning of the program accordingly.

(ii) You may need to make some changes to this program in order to make it run correctly on your computer. *See §3.4 for details.*

C02AEF specimen program

```
C       C02AEF: SOLUTION OF POLYNOMIAL EQUATION

        REAL A(10), ZRE(10), ZIM(10), TOL
        INTEGER N, IFAIL, NSOL, J

        TOL = 0.0

        WRITE (6,*)
      *  'ENTER THE NUMBER OF COEFFICIENTS IN THE POLYNOMIAL'
        READ (5,*) N
        NSOL = N - 1
        WRITE (6,*) 'ENTER THE COEFFICIENTS OF THE POLYNOMIAL'
        READ (5,*) (A(J),J=1,N)
```

```
      IFAIL = 0

      CALL C02AEF(A,N,ZRE,ZIM,TOL,IFAIL)

      WRITE (6,*) 'THE SOLUTIONS ARE'
      DO 10 J = 1, NSOL
        WRITE (6,*) ZRE(J), ' + ', ZIM(J), ' I'
   10 CONTINUE

      STOP

      END
```

C02AEF specimen run

```
ENTER THE NUMBER OF COEFFICIENTS IN THE POLYNOMIAL
6
ENTER THE COEFFICIENTS OF THE POLYNOMIAL
3.1   2.2   -7.3   -1.4   -5.8   -1.1
THE SOLUTIONS ARE
-1.992099 + 0.0000000E+00 I
1.553422 + 0.0000000E+00 I
-4.0798014E-02 + -0.7770020 I
-4.0798014E-02 + 0.7770020 I
-0.1894046 + 0.0000000E+00 I
```

Comments:

(i) In the specimen runs in this book, the information which you would need to type in at the terminal is reversed black to white. The reversed numbers typed in above are those which are needed to solve the specimen problem.

(ii) Most of the programs given in this book, including the one above, are written for 'on-line' use. In these programs, each time you are required to enter some data, the instruction to do this is accompanied by an input prompt. For example, in the program above, the pair of statements

```
WRITE (6,*)
* 'ENTER THE NUMBER OF COEFFICIENTS IN THE POLYNOMIAL'
READ (5,*) N
```

go together and should help you to enter the correct data at the right time.

If you are entering your data in any other way (such as from a data file or batch-mode) then these pairs of statements should be changed slightly. For instance, the pair of statements above could be replaced by

```
WRITE (6,*)
* 'THE NUMBER OF COEFFICIENTS IN THE POLYNOMIAL'
READ (5,*) N
WRITE (6,*) N
```

Thus, when the program is run, a suitable message will appear, followed by the data. It is good practice to get the data printed out, so that you can check that it has been correctly entered.

The program in §5.2 is a variation of the program above, adapted for use with a data file. Comparison of these two programs should give you sufficient guidance on how to adapt the other programs in this book if you need to.

(iii) You should be prepared to find that the layout and the accuracy of the results obtained on your computer differ slightly from those printed in this book. The following results were obtained on two different types of computer.

(a)
```
ENTER THE NUMBER OF COEFFICIENTS IN THE POLYNOMIAL
6
ENTER THE COEFFICIENTS OF THE POLYNOMIAL
3.1   2.2   -7.3   -1.4   -5.8   -1.1
THE SOLUTIONS ARE
  -1.992099        +     0.0000000E+00 I
   1.553422        +     0.0000000E+00 I
  -4.0798020E-02   +    -0.7770020     I
  -4.0798020E-02   +     0.7770020     I
  -0.1894046       +     0.0000000E+00 I
```

These results differ slightly from those given in the specimen run; note also that the layout is slightly different.

(b)
```
ENTER THE NUMBER OF COEFFICIENTS IN THE POLYNOMIAL
6
ENTER THE COEFFICIENTS OF THE POLYNOMIAL
3.1   2.2   -7.3   -1.4   -5.8   -1.1
THE SOLUTIONS ARE
  -1.99209922772634      +   0.000000000000000E+00 I
   1.55342242627535      +   0.000000000000000E+00 I
  -4.079801277136486E-02 +  -0.777001983255437     I
  -4.079801277136486E-02 +   0.777001983255437     I
  -0.189404592361112     +   0.000000000000000E+00 I
```

These results were obtained from a computer which works with about 15 decimal digits of precision, and prints 15 digits in the answers. You can see that when these are rounded to 7 digits they agree very closely with the original specimen run.

Postscript

Accuracy:
The routine evaluates the solutions as accurately as possible. However, the full number of significant figures printed in the specimen run should not be relied on. This is demonstrated by comparing the different results in the three specimen runs above.

Associated routines:
If you need to find the solutions of a polynomial equation where the coefficients $a_1, a_2, \ldots, a_n$ are complex, use routine C02ADF. Details of this routine can be found in the *NAG Manual*.

4.2 Solutions of $f(x) = 0$: C05ADF
   [An easy routine]

The **purpose** of this routine is to find a solution of the equation

$$f(x) = 0 \qquad\qquad (4.2)$$

in a given interval $[a,b]$, where f is a given function.

Specimen problem

To find the solution of the equation

$$e^{-x} - x = 0$$

which lies in the interval $[0,1]$.

The **method** uses a mixture of linear interpolation, extrapolation, and bisection.

The **routine name** with parameters is

C05ADF (A, B, EPS, ETA, F, X, IFAIL).

Description of parameters

Parameters which require values before C05ADF is called

$\left.\begin{array}{l} A \\ B \end{array}\right\}$: [real variables]

A and B must specify the interval $[a,b]$ in which you wish to find a solution.

Comments:
   (i) A must be less than B.
   (ii) The routine works only if f is continuous in the interval $[a,b]$ and if $f(a)$ and $f(b)$ have different signs (thus ensuring the existence of a root in $[a,b]$).

If you were looking for a root of $e^{-x} - x = 0$ in $[0,1]$,

$$f(x) = e^{-x} - x$$
so $f(0) = 1 \; (>0)$
and $f(1) = -0.63 \; (<0)$.

Thus, as $f(0)$ and $f(1)$ have different signs, and f is continuous, 0.0 and 1.0 would be suitable choices for A and B.

EPS: [a positive real variable]
The routine attempts to find an approximation X to an exact solution α of $f(x) = 0$. One of the stopping criteria is that the approximation X will be returned by the routine when

$$|X - \alpha| < EPS \qquad (EPS > 0).$$

So you have to supply a positive value for EPS, the absolute tolerance to which the solution is required.

Ref.
§5.4(b) Suppose, for instance that you wanted an answer which was likely to be accurate to 2 decimal places. Then you would choose a value EPS = 0.005

ETA: [a real variable]
Another stopping criterion for this routine is that the approximation X will be returned by the routine when

$$|f(X)| < \text{ETA} \qquad (\text{ETA} \geq 0).$$

So, you also have to supply a non-negative value for ETA.
  For instance if your purpose was to find a value of x for which $f(x)$ was within 0.01 of zero, then you should give a value ETA = 0.01.

Comment:
The approximation X is returned as soon as one of the criteria above is satisfied. In particular, if you give a value 0.0 for ETA, then the second criterion can never be satisfied. In this case, only the absolute tolerance criterion (EPS) will be applied.

Functions which require definition

F: [a real function. It must be declared as EXTERNAL at the beginning of your calling program]

Ref.
§2.2 F is the function f in the equation (4.2) above whose solution is required. You will have to supply a routine F(U) to define the function f.

For the specimen problem, a suitable routine could be written as follows:
```
REAL FUNCTION F(U)
REAL U
F = EXP (-U) -U
RETURN
END
```

Comment:
In the routine above, U is a parameter which gets its value from C05ADF when necessary. All you are required to do here is to define the function F.

The error parameter

IFAIL: [an integer variable]
IFAIL is the error parameter described in §3.2. It is recommended that you set

IFAIL = 0

before you call C05ADF. Then in the event of the routine failing, your program will stop and print one of the following messages:

Error message	Meaning	Advice
IFAIL = 1	*Either* A = B *or* F(A)*F(B) > 0 *or* EPS ⩽ 0.	Check that you have chosen suitable values for A, B, and EPS.
IFAIL = 2	Too much accuracy has been requested.	Try making EPS larger.
IFAIL = 3	There is some evidence that *f* has a discontinuity in [*a*,*b*].	Check your function for singularities in [*a*,*b*].
IFAIL = 4	A serious error has occurred.	Check your CALL statement. Get some help.

Comment:
If in any doubt as to what to do when the routine fails, seek some expert help.

Parameters to be examined after calling C05ADF

X: [a real variable]
Assuming everything has gone well, X will contain the final approximation to a solution of the equation $f(x) = 0$ which lies in the interval [*a*,*b*].

Comment:
The routine C05ADF finds only one solution in [*a*,*b*], even if there were more solutions in this interval. It is up to the user to check that the interval [*a*,*b*] is sufficiently small to contain only the solution required.

Specimen program

Program planning

1. *Declare*   REAL A, B, EPS, ETA, X
                 INTEGER IFAIL
                 EXTERNAL F

2. *Read*   A, B, EPS, ETA
   Set   IFAIL

3. *Call*   C05ADF

4. *Print*   X

5. *Write routine* F(U)

Comments:

(i) The following program specifically finds a solution of $e^{-x} - x = 0$ in [0,1]. When solutions are required for other equations, then you will have to change the function F accordingly. You will also have to find a suitable interval $[a,b]$ in which there is a solution.

(ii) You may need to make some changes to this program in order to make it run correctly on your computer. *See §3.4 for details.*

C05ADF specimen program

```
C       C05ADF: SOLUTION OF F(X)=0

        REAL A, B, EPS, ETA, X
        INTEGER IFAIL
        EXTERNAL F

        WRITE (6,*) 'ENTER THE VALUES OF A AND B'
        READ (5,*) A, B
        WRITE (6,*) 'ENTER THE VALUES OF EPS AND ETA'
        READ (5,*) EPS, ETA

        IFAIL = 0

        CALL C05ADF(A,B,EPS,ETA,F,X,IFAIL)

        WRITE (6,*) 'APPROXIMATION TO SOLUTION IS ', X

        STOP

        END

        REAL FUNCTION F(U)
        REAL U
        F = EXP(-U) - U
        RETURN
        END
```

C05ADF specimen run 1

```
ENTER THE VALUES OF A AND B
 0.0   1.0
ENTER THE VALUES OF EPS AND ETA
 0.5E-4   0.0
APPROXIMATION TO SOLUTION IS  0.5671 433
```

Postscript

Accuracy:

Note that in the program run above, a value 0.5E−4 (0.00005) was supplied for EPS. In these circumstances, only the 4 decimal places boxed above are likely to be accurate in the answer.

Comment:

If you mistakenly specify an interval $[a,b]$ where there is no root, then the routine would stop and print its own error message as demonstrated below:

C05ADF specimen run 2

```
ENTER THE VALUES OF A AND B
1.0    2.0
ENTER THE VALUES OF EPS AND ETA
0.5E-4   0.0

ERROR DETECTED BY NAG LIBRARY ROUTINE   C05ADF   - IFAIL =    1
```

Associated routines:

(i) If your equation is a polynomial equation, then use C02AEF, the routine described in §4.1.

(ii) If you have simultaneous non-linear equations, then use C05NBF. You will find the documentation for this in §10.4.

4.3 Linear simultaneous equations: F04ARF [An easy/medium routine]

The **purpose** of this routine is to find the solution of a set of n linear simultaneous equations in n unknowns.

Comment:

In matrix terms, the routine finds the solution $\mathbf{x}$ of the matrix equation

$$\mathbf{Ax} = \mathbf{b} . \tag{4.3}$$

Specimen problem

To solve the equations
$$
\begin{aligned}
33x_1 + 16x_2 + 72x_3 &= -359 \\
-24x_1 - 10x_2 - 57x_3 &= 281 \\
- 8x_1 - 4x_2 - 17x_3 &= 85
\end{aligned}
$$

Comment:

Equations like the ones above can always be written as a matrix equation $A\mathbf{x} = \mathbf{b}$. In particular, these equations can be rewritten as

$$
\begin{bmatrix} 33 & 16 & 72 \\ -24 & -10 & -57 \\ -8 & -4 & -17 \end{bmatrix} \begin{bmatrix} x_1 \\ x_2 \\ x_3 \end{bmatrix} = \begin{bmatrix} -359 \\ 281 \\ 85 \end{bmatrix}
$$

The **method** first decomposes A using Crout's factorization, and then uses forward and back-substitution. Further information about the Crout factorization can be found in the *NAG Manual*, if required.

The **routine name** with parameters is

F04ARF (A, IA, B, N, X, WKSPCE, IFAIL).

Description of parameters

Parameters which require values before F04ARF is called

N: [an integer variable]
N must contain the number of equations to be solved.

In the case of the specimen problem, there are three equations, so N should be given the value 3.

A: [a real two-dimensional array. This array should have at least N rows and at least N columns in the REAL declaration in the calling program]
The array A is used to contain the coefficients of the left-hand sides of your equations.

Ref.
p. 19

Thus, in the case of the specimen problem

$$A(1,1) = 33.0 \quad A(1,2) = 16.0 \quad A(1,3) = 72.0$$
$$A(2,1) = -24.0 \quad \ldots \text{ and so on.}$$

IA: [an integer variable]
IA must specify the size of the first dimension of A as declared in your calling program.

Ref.
§2.4

So, if you had a (not very realistic) declaration

REAL A(10,20)

in your calling program, then the statement

IA = 10

would also be needed.

B: [a real one-dimensional array. Its length should be at least N in the REAL declaration in your calling program]
The array B should contain the right-hand sides of the equations.

In the case of the specimen problem,

$$B(1) = -359.0 \quad B(2) = 281.0 \quad \text{and} \quad B(3) = 85.0.$$

Parameters associated with workspace

WKSPCE: [a real one-dimensional array. Its length must be at least N in the REAL declaration in your calling program]
This array is used as workspace.

The error parameter

IFAIL: [an integer variable]
IFAIL is the error parameter described in §3.2. It is recommended that you set

IFAIL = 0

before you call F04ARF. Then in the event of the routine failing, your program will stop and print

IFAIL = 1.

This means that the matrix A is singular (possibly due to rounding error). In this event, get some expert help.

Parameters to be examined after calling F04ARF

X: [a real one-dimensional array. Its length must be at least N in the REAL declaration in your calling program]
If all goes well in the routine, then the solution $x_1, x_2, \ldots, x_n$ of the equations will be found in X(1), X(2), . . ., X(N) respectively.

Specimen program

Program planning

1. *Declare* REAL A(,), B( ), X( ), WKSPCE( )
   INTEGER IA, N, IFAIL

2. *Set*   IA
   Read   N, A, B
   Set   IFAIL

3. *Call*   F04ARF

4. *Print*   X

Comments:
    (i) The following program can be used to solve sets of up to 10 equations in 10 unknowns. If you want to use the program for more equations, then you will have to alter the dimensions of the arrays in the REAL declaration accordingly. You will also have to change the value given to IA in the program.
    (ii) You may need to make some changes to this program in order to make it run correctly on your computer. *See §3.4 for details.*

F04ARF specimen program

```
C        F04ARF: LINEAR SIMULTANEOUS EQUATIONS

         REAL A(10,10), B(10), X(10), WKSPCE(10)
         INTEGER IA, N, IFAIL, I, J

         IA = 10

         WRITE (6,*) 'ENTER THE NUMBER OF EQUATIONS'
         READ (5,*) N
         WRITE (6,*) 'ENTER THE MATRIX A'
         DO 10 I = 1, N
           READ (5,*) (A(I,J),J=1,N)
      10 CONTINUE
         WRITE (6,*)
       *   'ENTER THE RIGHT-HAND SIDE COEFFICIENTS, ',
       *   'ONE PER LINE'
         DO 20 I = 1, N
           READ (5,*) B(I)
      20 CONTINUE

         IFAIL = 0

         CALL F04ARF(A,IA,B,N,X,WKSPCE,IFAIL)

         WRITE (6,*) 'THE SOLUTION IS'
         DO 30 I = 1, N
           WRITE (6,*) X(I)
      30 CONTINUE

         STOP

         END
```

F04ARF specimen run

```
ENTER THE NUMBER OF EQUATIONS
 3
ENTER THE MATRIX A
 33.0   16.0   72.0
-24.0  -10.0  -57.0
 -8.0   -4.0  -17.0
ENTER THE RIGHT-HAND SIDE COEFFICIENTS, ONE PER LINE
-359.0
 281.0
  85.0
THE SOLUTION IS
1.000000
-2.000000
-5.000000
```

Postscript

Accuracy:

The results are likely to be reasonably accurate so long as your equations are well-conditioned.

Comment:

Ref.
§5.2
If your matrix A is large, then you are recommended to use a data file.

Associated routines:

(i) A more general routine is F04AEF which is described in §6.10. This allows several right-hand sides to be processed in a single call of the routine.

(ii) If A is a large symmetric positive-definite matrix, then use ,F04ASF. You will find the documentation for this in §9.

(iii) If A is a large positive-definite matrix which has band structure, then use F04ACF which you will find in §10.6.

(iv) If your equations contain complex numbers, then use F04ADF which you will find in §10.3.

(v) If you are often likely to be using routines to solve simultaneous equations, or if your matrix has special properties (such as the ones described above), then you should consult the appropriate decision tree in the *NAG Mini-Manual*. This will help you to choose the routine in the *NAG Manual* best suited to your problem. You will find this particular decision tree in §8.

5
Avoiding mistakes

To people who are new to the computer and numerical scene, the very fact that the computer has printed a number is greeted with great relief. However, this relief does not necessarily mean that the number which is printed is the correct answer to the problem. The number or so-called 'answer' must be treated with the greatest of caution.

This chapter aims to help you to avoid making mistakes, and to give you ideas about checking the accuracy or validity of the answer which you have found. If your answer or results are to form the basis of further work, it is, of course, of great importance to test extensively the validity of the results. The routines chosen for demonstration purposes in this chapter are C02AEF and C05ADF, as it is assumed that most readers will have a working knowledge of these routines at this stage.

The basic specimen programs for these routines can be found in §4.1 and §4.2.

5.1 Preliminary runs

Once you have understood how to use one of the NAG routines, don't be in too much haste to run your own problem. The following precautions are the *minimum* which you should take before you start to have any faith in the answers obtained from your program.

(a) Run your program using the data given for the specimen problem in the description of the NAG routine. Check that you get the same answer.

(b) Run your program again, using data from a problem whose answer you know. Check that you get the answer you expect.

(c) If your program uses more than one NAG routine, check each part of your program separately to ensure that each routine is producing the expected answer. Then test the whole program.

Only after taking these precautions would it be reasonable to consider using the program to solve your own problem.

5.2 Data checking and data files

It is very easy to make mistakes when typing in data. In Chapter 4, the specimen programs were written so that you typed in the data when you wanted to run the program. This, in fact, is an entirely unsuitable way of entering large quantities of data into the machine. If even one mistake is made (and not detected immediately), then you would be forced to type in all the data again. In a problem where there is a lot of data involved, it is good practice to put your data in a *data file*.

A data file can be edited in exactly the same way as any other file. Thus, if you make a mistake when typing in the data, you will be able to correct it.

If you are going to use a data file, then you may need an 'OPEN' statement at the beginning of your program. This informs the computer

(a) what the name of the data file is, and
(b) how it can be accessed.

A typical 'OPEN' statement would be

OPEN (UNIT = 20, FILE = 'MAT.DAT') .

This statement carries the information that the data in the file MAT.DAT can be accessed via unit number 20. Thus, when the computer meets the instruction

READ (20,*) N ,

then it reads the value of N from the data file MAT.DAT.

More details on using a data file can be found in a Fortran manual. Your computer centre will be able to advise you how to use data files on your machine.

The following specimen program for C02AEF is a variation of the one given in §4.1. This time the coefficients of the equation are read from a data file C02AEF.DAT, which contains data arranged in the following way:

```
6
3.1 2.2 -7.3 -1.4 -5.8 -1.1
```

Specimen program using a data file

```
C       C02AEF: SOLUTION OF POLYNOMIAL EQUATION
C               (USING DATA FILE)

        REAL A(10), ZRE(10), ZIM(10), TOL
        INTEGER N, IFAIL, NSOL, J

        OPEN (UNIT=20,FILE='C02AEF.DAT')

        TOL = 0.0
```

continued

```
        WRITE (6,*)
      *   'THE NUMBER OF COEFFICIENTS IN THE POLYNOMIAL'
        READ (20,*) N
        WRITE (6,*) N
        NSOL = N - 1
        WRITE (6,*) 'THE COEFFICIENTS OF THE POLYNOMIAL'
        READ (20,*) (A(J),J=1,N)
        WRITE (6,*) (A(J),J=1,N)

        IFAIL = 0

        CALL C02AEF(A,N,ZRE,ZIM,TOL,IFAIL)

        WRITE (6,*) 'THE SOLUTIONS ARE'
        DO 10 J = 1, NSOL
           WRITE (6,*) ZRE(J), ' + ', ZIM(J), ' I'
     10 CONTINUE

        STOP

        END
```

A specimen run of this program is given below:

```
THE NUMBER OF COEFFICIENTS IN THE POLYNOMIAL
6
THE COEFFICIENTS OF THE POLYNOMIAL
3.10000, 2.20000, -7.30000, -1.40000, -5.80000, -1.10000
THE SOLUTIONS ARE
-1.992099 + 0.0000000E+00 I
1.553422 + 0.0000000E+00 I
-4.0798014E-02 + -0.7770020 I
-4.0798014E-02 + 0.7770020 I
-0.1894046 + 0.0000000E+00 I
```

Comment:
One major source of error in a problem is entering incorrect data. Incorrect data may well produce an answer, but it won't be the answer to your intended problem. So it is good programming practice to print out the original data along with the answer. In some cases, a graphical display of the data might well show up some error.

5.3 Further use of IFAIL

In the specimen programs in Chapter 4, very limited use was made of IFAIL. The effect of setting

IFAIL = 0

initially was that the program would stop if anything went wrong. In this event, the routine printed its own error messages.

The other standard way of using IFAIL is to set

IFAIL = 1

initially. In most NAG routines, if you do this and anything goes wrong, then control is passed back to your program *without* an error message being printed. In other words, the program goes on, irrespective of the value of IFAIL. So, if you set IFAIL = 1 initially, you *must* test or print the value of IFAIL *immediately* after you have called the routine to check if any error has been detected. If IFAIL = 0 after a call of a routine, then no error has been detected. However, if IFAIL has any other value after a call of a routine, then you should check what went wrong.

It is not always necessary or convenient to re-run a program from scratch each time an error message crops up. To demonstrate this in a simple case, look at the adaptation of C05ADF given below.

Specimen program using IFAIL = 1

```
C       C05ADF: SOLUTION OF F(X)=0

        REAL A, B, EPS, ETA, X
        INTEGER IFAIL
        EXTERNAL F

     10 WRITE (6,*) 'ENTER THE VALUES OF A AND B'
        READ (5,*) A, B
        WRITE (6,*) 'ENTER THE VALUES OF EPS AND ETA'
        READ (5,*) EPS, ETA

        IFAIL = 1

        CALL C05ADF(A,B,EPS,ETA,F,X,IFAIL)

        IF (IFAIL.EQ.0) THEN
          WRITE (6,*) 'APPROXIMATION TO SOLUTION IS ', X
        ELSE IF (IFAIL.EQ.1) THEN
          WRITE (6,*)
     *      'THERE IS NO SOLUTION IN SPECIFIED INTERVAL'
          WRITE (6,*) 'OR EPS .LE. 0.0 - TRY AGAIN'
          GO TO 10
        ELSE
          WRITE (6,*) 'IFAIL =', IFAIL
        END IF

        STOP

        END

        REAL FUNCTION F(U)
        REAL U
        F = EXP(-U) - U
        RETURN
        END
```

continued

A run of this program is given below

```
ENTER THE VALUES OF A AND B
 1.0   2.0
ENTER THE VALUES OF EPS AND ETA
 0.5E-4   0.0
THERE IS NO SOLUTION IN SPECIFIED INTERVAL
OR EPS .LE. 0.0 - TRY AGAIN
ENTER THE VALUES OF A AND B
 0.0   1.0
ENTER THE VALUES OF EPS AND ETA
 0.5E-4   0.0
APPROXIMATION TO SOLUTION IS 0.5671433
```

5.4 Accuracy of the answer

There are various distinct factors which control how many significant figures can be relied on in a given situation. They are treated separately here. However, when you are actually solving a problem, they all have to be taken into account.

(a) *Programming note*

When using a NAG routine to solve a problem, there is often no prior knowledge to tell you how much accuracy a particular run is likely to produce. So it is impossible to write a program which prints an answer correct to a guaranteed number of significant figures.

   In the specimen programs, most of the answers are printed using a Fortran 77 statement such as

   WRITE (6,*) A .

This statement normally prints A to the maximum number of significant figures that is likely to be accurate in the computer, assuming that everything else was entirely accurate. However, this is virtually never the case. For instance, any problem involving a substantial amount of computation will introduce rounding error which will affect the answer. So, it would be extremely foolish to rely on all the figures which are printed. The main factors which influence how many significant figures on which to rely are discussed under (b) and (c) below.

(b) *Mathematical note*

Normally in an iterative process, if an answer is required to k decimal places, then it is good enough to stop the process when the absolute accuracy test

$$|x_{n+1} - x_n| < \tfrac{1}{2} \times 10^{-k}$$

is satisfied. This does not actually guarantee an answer correct to k decimal places, but as a rule of thumb, it generally works.

Using a similar rule of thumb, if an answer is required to k significant figures, then this is usually achieved when the relative accuracy test

$$\left| \frac{x_{n+1} - x_n}{x_n} \right| < \tfrac{1}{2} \times 10^{-k}$$

is satisfied.

In some NAG routines, you are asked to supply stopping criteria for either the absolute or relative accuracy which you require in your answer. Although you are asked to supply this information, the routine often does not guarantee to supply the answer to your requested accuracy. In this case you should certainly run your program more than once, with different stopping criteria. Then compare the answers you get using the different tolerances, and see to how many figures the different answers agree. The figures which agree are the only ones which it would be reasonably safe to quote in an answer.

The program below is a simple adaptation of the program given in §4.2 with a loop to pick up two sets of tolerances.

```
C       C05ADF: SOLUTION OF F(X)=0
C               (WITH ACCURACY LOOP)

        REAL A, B, EPS, ETA, X
        INTEGER IFAIL, I
        EXTERNAL F

        WRITE (6,*) 'ENTER THE VALUES OF A AND B'
        READ (5,*) A, B
        DO 10 I = 1, 2
          WRITE (6,*) 'ENTER THE VALUES OF EPS AND ETA'
          READ (5,*) EPS, ETA

          IFAIL = 0

          CALL C05ADF(A,B,EPS,ETA,F,X,IFAIL)

          WRITE (6,*) 'APPROXIMATION TO SOLUTION IS ', X
     10 CONTINUE

        STOP

        END

        REAL FUNCTION F(U)
        REAL U
        F = EXP(-U) - U
        RETURN
        END
```

Using absolute tolerances of $\tfrac{1}{2} \times 10^{-3}$ and $\tfrac{1}{2} \times 10^{-4}$, the following results were produced.

```
ENTER THE VALUES OF A AND B
 0.0  1.0
ENTER THE VALUES OF EPS AND ETA
 0.5E-3  0.0
APPROXIMATION TO SOLUTION IS 0.5671704
ENTER THE VALUES OF EPS AND ETA
 0.5E-4  0.0
APPROXIMATION TO SOLUTION IS 0.5671433
```

Now, as the two answers agree to 3 significant figures, it is reasonable to quote 0.567 as an answer.

(c) *Physical note*

Suppose the data you have for your problem are given to three significant figures. Then, however much accuracy you ask for in your answer, and however many significant figures you choose to print, it is unlikely that the answer printed will be accurate to more than three significant figures. In cases where a complicated calculation has induced significant rounding error, not even three significant figures would be correct. So, don't expect more accuracy in your output than you had in your input.

5.5 Reliability of the answer (ill-conditioning)

A problem is said to be ill-conditioned if small changes in the data produce significant changes in the solution.

Ill-conditioning is, perhaps, the most serious problem discussed in this chapter. It is often difficult to detect, and even more difficult to remedy.

The reason why this problem is so serious is that, when a mathematical model is formed from a real-life situation, the data used are seldom, if ever, exact. Take for example a ruler measurement, or a population count. A measurement 6.2 cm on a ruler may cover all measurements from 6.15 cm to 6.25 cm, and a population count of 6000 could very easily be 5999 or 6001, in fact. As, in practice, very few problems can be formulated exactly, it is hoped that these inevitable small changes will produce very little difference in the solutions obtained. Unfortunately, this is not always the case.

The moral tale below produces an ill-conditioned set of linear simultaneous equations. Note, however, that ill-conditioning can be found in many other types of problems, notably in differential equations, and non-linear equations.

Tale:

John ordered 6 large loaves, 4 chocolate cakes, and 3 seed cakes from a baker, and got a bill for £7.34. Now, not knowing the individual prices of the items, he assumed that

$$6b + 4c + 3s = 734 \, ,$$

where b, c, and s are the prices (in pence) of a loaf, a chocolate cake, and a seed cake respectively.

From two more orders he obtained two more equations

$$7\tfrac{1}{2}b + 6c + 5s = 1049\tfrac{1}{2}$$
and $\quad 10b + 7\tfrac{1}{2}c + 6s = 1331 \, .$

These three equations have an exact solution $b = 53$, $c = 62$ and $s = 56$, and so John assumed that he knew the price of the items without asking the baker.

Now, John's friend Joyce performed the same exercise. The only difference was that Joyce was served by the baker's assistant, who charges an extra $\tfrac{1}{2}$p for every half loaf he sells. So, as the first order included two half-loaves and the second order included five half-loaves, the first and second order cost 1p and 2$\tfrac{1}{2}$p more than John's orders. So, Joyce found herself with the equations

$$6b + 4c + 3s = 735$$
$$7\tfrac{1}{2}b + 6c + 5s = 1052$$
and $\quad 10b + 7\tfrac{1}{2}c + 6s = 1331$

which she expected to have roughly the same solutions as John's equations. Much to the consternation of both John and Joyce, these slightly changed equations have an exact solution $b = 74$, $c = -18$ and $s = 121$.

Discussion:

(i) The trouble in the tale above is that the equations are ill-conditioned. A small (reasonable) change in the data produces an answer which bears no resemblance to the 'real' solution to the problem.

(ii) Note that the ill-conditioning lies in the equations themselves, and not in the method (or NAG routines) used to solve them.

(iii) Even if a problem can be formulated exactly, the process of solving it on a computer introduces errors which can have a similar effect to small changes in the data. If a problem is ill-conditioned, the computed results may differ drastically from the exact results.

Advice:

(i) Unlike the situation in the tale above, normally there is just one problem to solve, so it is always worth trying to check that a problem is well-conditioned. Take the trouble to solve some 'neighbouring' problems, with slightly changed data, to check whether the resulting solutions differ significantly from the original ones.

(ii) If you find, or suspect, that your problem is ill-conditioned, then you should be extremely cautious about the answer. Try to find an alternative formulation of the problem which is not ill-conditioned.

If in doubt as to what to do, get some expert help or advice.

Part 2
Some 'easy-to-use' NAG routines

This part of the book consists of descriptions of a selection of routines from the *NAG Library*. The descriptions given here and the specimen programs should be easier to use than those given in the *NAG Manual*.

You are not expected to work through this section in order. Select the routines that you want to use from this section. However, first check that you understand the contents of Part 1.

Part 2 also includes some NAG graphical routines. Make certain that you read the Introduction to the Graphical Chapter before you attempt to use these routines.

6
Routine descriptions

6.1 Discrete Fourier Transform: C06ECF
[An easy routine]

The **purpose** of this routine is to calculate the discrete Fourier Transform of n complex data values $z_1, z_2, \ldots, z_n$.

Comments:

(i) The definition of the discrete Fourier Transform used in this routine is

$$\hat{z}_k = \frac{1}{\sqrt{n}} \sum_{m=0}^{n-1} z_m \times \exp\left\{ -i \; \frac{2\pi mk}{n} \right\} \tag{6.1}$$

where $k = 0, 1, \ldots, n - 1$, and $i = \sqrt{-1}$.

(ii) The scale factor $(1/\sqrt{n})$ given here may not be suitable for all applications. If you require a different scale factor, you can adjust the specimen program to multiply the results by whatever factor you need.

(iii) The Fourier Transform of a function $f(t)$ indicates the relative importance of the different frequencies which go to make up $f(t)$. You should consult a book on Fourier Analysis to understand the relationship between the discrete transform computed here, and the continuous transform which is used in the theory of Fourier Analysis.

Specimen problem

To find the discrete Fourier Transform of the 4 data values

$2 + 3i, 3 - i, 7 + 3i, 1 + i$.

Comment:

In any practical problem, there would be many more data values than this.

The **method** uses the Fast Fourier Transform algorithm.

The **routine name** with parameters is

C06ECF (X, Y, N, IFAIL) .

Description of parameters

Parameters which require values before C06ECF is called

N: [an integer variable]
N is the number of data values.

In the specimen problem there are 4 data values, so N should be specified as 4.

Comments:
(i) N must be greater than 1.
(ii) The largest prime factor N is allowed to have is 19.
(iii) The maximum number of prime factors N may have, including repetitions, is 20.

X $\Big\rbrace$. [real one-dimensional arrays. Both arrays must have length at least N in
Y $\Big\rbrace$ the REAL declaration in your calling program]

Ref. X(J) and Y(J) are used to contain the real and imaginary parts
p. 19 respectively of the j(th) data-value (denoted by z_{j-1} in (6.1) above).

Thus, using the specimen problem as an example

$$X(1) = 2.0, \quad Y(1) = 3.0$$
$$X(2) = 3.0, \quad Y(2) = -1.0$$

and so on.

The error parameter

IFAIL: [an integer variable]
IFAIL is the error parameter described in §3.2. It is recommended that you set

$$IFAIL = 0$$

before you call C06ECF. Then in the event of the routine failing, your program will stop and print one of the following error messages:

Error message	Meaning	Advice
IFAIL = 1	One of the prime factors of N is greater than 19.	In all cases, check the restrictions on N, and rectify the error.
IFAIL = 2	N has more than 20 factors.	
IFAIL = 3	N ≤ 1.	

Parameters to be examined after calling C06ECF

$\left.\begin{array}{l}X\\Y\end{array}\right\}$: [the two one-dimensional arrays described earlier]

Originally, the arrays X and Y were used to hold the N data values. However, after the routine has been called, the arrays X and Y will hold the real and imaginary parts of the N components of the discrete Fourier Transform. In particular, X(J) and Y(J) will contain respectively the real and imaginary part of the j(th) component (denoted by $\hat{z}_{j-1}$ in (6.1) above).

Specimen program

Program planning

1. *Declare* REAL X( ), Y( )
   INTEGER N, IFAIL

2. *Read* N, X, Y
   Set IFAIL

3. *Call* C06ECF

4. *Print* X, Y

Comments:

(i) The following program will obtain the components of the discrete Fourier Transform for up to 100 data values. If you have more data values than this, then you will have to alter the REAL declarations at the beginning of the program accordingly.

(ii) You may need to make some changes to this program in order to make it run correctly on your computer. *See §3.4 for details.*

C06ECF specimen program

```
C     C06ECF: DISCRETE FOURIER TRANSFORM

      REAL X(100), Y(100)
      INTEGER N, IFAIL, J

      WRITE (6,*) 'ENTER THE NUMBER OF DATA VALUES'
      READ (5,*) N
      WRITE (6,*)
    * 'ENTER THE (COMPLEX) DATA VALUES, ONE PER LINE'
      WRITE (6,*) 'REAL PART  IMAGINARY PART'
      DO 10 J = 1, N
        READ (5,*) X(J), Y(J)
   10 CONTINUE

      IFAIL = 0

      CALL C06ECF(X,Y,N,IFAIL)
```

continued

```
WRITE (6,*) 'COMPONENTS OF DISCRETE FOURIER TRANSFORM'
WRITE (6,*) 'REAL PART   IMAGINARY PART'
DO 20 J = 1, N
   WRITE (6,*) X(J), Y(J)
20 CONTINUE

STOP

END
```

C06ECF specimen run

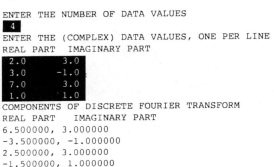

```
ENTER THE NUMBER OF DATA VALUES
4
ENTER THE (COMPLEX) DATA VALUES, ONE PER LINE
REAL PART   IMAGINARY PART
2.0        3.0
3.0       -1.0
7.0        3.0
1.0        1.0
COMPONENTS OF DISCRETE FOURIER TRANSFORM
REAL PART   IMAGINARY PART
6.500000, 3.000000
-3.500000, -1.000000
2.500000, 3.000000
-1.500000, 1.000000
```

6.2 Integral of a function: D01AJF
[A medium routine]

The **purpose** of this routine is, given a function f, to calculate an approximation to the integral

$$I = \int_a^b f(x)\,dx \ . \tag{6.2}$$

Specimen problem

To evaluate the integral

$$I = \int_{0.5}^1 \frac{1 - e^{-x}}{x} \ dx \ .$$

The **method** is an adaptive algorithm based on the Gauss 10-point and Kronrod 21-point rules.

The **routine name** with parameters is

D01AJF (F, A, B, EPSABS, EPSREL, RESULT, ABSERR, W, LW, IW, LIW, IFAIL).

Description of parameters

Parameters which require values before D01AJF is called

$\left.\begin{array}{l} A \\ B \end{array}\right\}$: [real variables]

A and B should contain the lower and upper limits of the integral I in (6.2). So, in the specimen problem, A = 0.5 and B = 1.0

EPSABS: [a real variable]

Ref.
§5.4(b)

EPSABS should specify the absolute accuracy which you would like in your answer. So, if you wanted an answer which was likely to be accurate to 3 decimal places, then you could choose a value EPSABS = 0.0005.

EPSREL: [a real variable]

EPSREL should specify the relative accuracy which you would like in your answer. So, if you wanted an answer which was likely to be accurate to 4 significant figures, then you could choose a value EPSREL = 0.00005.

Comments:

(i) Although you are asked to give your accuracy requirements, the routine does not guarantee to satisfy either of them exactly. An estimate of the actual accuracy achieved can be found in ABSERR which is returned after a call to the routine.

(ii) If you are concerned only with the relative accuracy in your answer, then you should put EPSABS = 0.0. Similarly, if you are only concerned with the absolute accuracy, then you should put EPSREL = 0.0.

[For further details see the *NAG Manual*.]

Functions which require definition

Ref.
§2.2

F: [a real function. It must be declared as EXTERNAL at the beginning of your program]

F is the function *f* in (6.2) above whose integral is required. You will have to supply a routine F(X) to define this function.

For the specimen problem, a suitable routine could be written as follows:

```
REAL FUNCTION F(X)
REAL X
F = (1.0 - EXP (-X))/X
RETURN
END
```

Comment:

In the routine above, X is a parameter which gets its value from D01AJF when necessary. All you are required to do here is to define the function F.

Parameters associated with workspace

W: [a real one-dimensional array. Its length must be at least 4 in the REAL

Ref.
§3.3
declaration in your calling program]
This array is used as workspace.

Comment:
Despite the trivial requirement stated above, most problems need a much larger W array. Basically, the more difficult the function is to integrate, the more workspace is needed. So, in the first instance, try a declaration

REAL W(400) .

If this is not enough, then a suitable error message (IFAIL = 1) will appear, and you can change the declaration accordingly.

LW: [an integer variable]

Ref.
§2.4
LW must specify the length of W as declared in your calling program. So, if you used the REAL declaration above, then a statement

LW = 400

would be needed in your program.

IW: [an integer one-dimensional array. Its length must be at least (LW/8 + 2) in the INTEGER declaration in your calling program]
This array is used as workspace.

Comment:
Once you have made a declaration of W, the size of the array IW is connected to that of W. For instance, if LW = 400, then

LW/8 + 2 = 400/8 + 2 = 52.

Thus, in this instance, a minimum declaration

INTEGER IW(52)

would be needed in your program.

LIW: [an integer variable]
LIW must specify the length of IW as declared in your calling program. So, if you used the INTEGER declaration above, then a statement

LIW = 52

would also be needed in your program.

The error parameter

IFAIL: [an integer variable]

IFAIL is the error parameter described in §3.2. It is recommended that you set

IFAIL = 0

before you call D01AJF. Then in the event of the routine failing, your program will stop and print one of the following messages:

Error message	Meaning	Advice
IFAIL = 1	Your problem is too difficult for the workspace allowed. Maybe a peak or discontinuity is causing trouble.	Alternative possibilities (a) See if you can find a point c where there is a peak or discontinuity. If so, use the routine twice on the intervals $[a,b]$ and $[c,b]$. (b) Allow more workspace. (c) Relax the EPSABS and EPSREL requirements. (d) Look for a routine better suited to your integral.
IFAIL = 2 IFAIL = 4	The accuracy you have requested is not being achieved.	Same advice as for IFAIL = 1 (c)
IFAIL = 3	The function is probably badly behaved somewhere.	Same advice as for IFAIL = 1 (a)
IFAIL = 5	*Either* the integral is diverging *or* the process is converging too slowly.	Seek expert help.
IFAIL = 6	*Either* LW < 4 *or* LIW < LW/8 + 2	Check your declaration of workspace, etc., carefully.

Comment:

If the integral is divergent, then any one of the IFAIL messages above may occur. So, if in any doubt as to what to do when the routine fails, seek some expert help.

Parameters to be examined after calling D01AJF

RESULT: [a real variable]

If everything has gone well, then RESULT will contain the required approximation to the integral, I, defined in (6.2).

ABSERR: [a real variable]
ABSERR will contain an estimate of the modulus of the absolute error to be expected in RESULT. Hence, you can expect

$$|\text{RESULT} - \text{I}| \leqslant \text{ABSERR}$$

Specimen program

Program planning

1. *Declare*   REAL A, B, EPSABS, EPSREL, RESULT, ABSERR, W( ),
   INTEGER LW, IW( ), LIW, IFAIL
   EXTERNAL F

2. *Read*   A, B, EPSABS, EPSREL
   Set   LW, LIW, IFAIL

3. *Call*   D01AJF

4. *Print*   RESULT, ABSERR

5. *Write routine* F(X)

Comments:

(i) The following program specifically finds an approximation for the integral in the specimen problem. When you want to use the program for other integrals, then you will have to change the function F accordingly. If necessary, you may also need to change the workspace declarations at the beginning of the program. In which case, you would also have to change the values given to LW and LIW.

(ii) You may need to make some changes to this program in order to make it run correctly on your computer. *See §3.4 for details.*

D01AJF specimen program

```
C      D01AJF: INTEGRAL OF A FUNCTION

       REAL A, B, EPSABS, EPSREL, RESULT, ABSERR, W(1200)
       INTEGER LW, IW(152), LIW, IFAIL
       EXTERNAL F

       LW = 1200
       LIW = 152

       WRITE (6,*) 'ENTER THE VALUES OF A AND B'
       READ (5,*) A, B
       WRITE (6,*) 'ENTER THE ABSOLUTE ACCURACY REQUIRED'
       READ (5,*) EPSABS
       WRITE (6,*) 'ENTER THE RELATIVE ACCURACY REQUIRED'
       READ (5,*) EPSREL
```

```
      IFAIL = 0

      CALL D01AJF(F,A,B,EPSABS,EPSREL,RESULT,ABSERR,W,LW,IW,
     *            LIW,IFAIL)

      WRITE (6,*) 'AN APPROXIMATION TO THE INTEGRAL IS ',
     * RESULT
      WRITE (6,*) 'WITH AN ESTIMATED ERROR OF ', ABSERR

      STOP

      END

      REAL FUNCTION F(X)
      REAL X
      F = (1.0-EXP(-X))/X
      RETURN
      END
```

D01AJF specimen run

```
ENTER THE VALUES OF A AND B
0.5  1.0
ENTER THE ABSOLUTE ACCURACY REQUIRED
0.0
ENTER THE RELATIVE ACCURACY REQUIRED
0.5E-4
AN APPROXIMATION TO THE INTEGRAL IS 0.3527575
WITH AN ESTIMATED ERROR OF 1.3141242E-07
```

Postscript

Accuracy:
Although only 4-significant figure accuracy was effectively requested
(0.5E$-$4) the estimated error was very small. So probably most of the
significant figures given in the answer would be reliable.

Associated routines:
   (i) If you need a routine which evaluates an integral from given data
points, then use D01GAF described on the next page.
   (ii) A multiple integral routine D01GBF is included in §10.5.
Ref. §8   (iii) If you often use integration routines, you would be advised to look
at the appropriate *decision tree* in the *NAG Mini-Manual* first. This will
help you choose the routine in the *NAG Manual* best suited to your
problem.

6.3 Integral from data points: D01GAF [An easy routine]

The **purpose** of this routine is to find an approximation to the integral

$$I = \int_{x_n}^{x_1} f(x)\mathrm{d}x \tag{6.3}$$

where the function $y = f(x)$ is specified at the data points (x_1, y_1), (x_2, y_2), . . ., (x_n, y_n).

Specimen problem

To find

$$\int_0^1 f(x)\mathrm{d}x$$

where the 9 following tabular values are known

k	x_k	$y_k = f(x_k)$
1	0.00	4.0000
2	0.08	3.9746
3	0.22	3.8153
4	0.39	3.4719
5	0.46	3.3014
6	0.60	2.9412
7	0.73	2.6094
8	0.85	2.3222
9	1.00	2.0000

The **method** uses a four-point finite difference formula in an algorithm developed by Gill and Miller.

The **routine name** with parameters is

D01GAF (X, Y, N, ANS, ER, IFAIL).

Description of parameters

Parameters which require values before D01GAF is called

N: [an integer variable]
N is the number of data points.

In the specimen problem, there are 9 data points, so N should be specified as 9 in this instance.

Comment:
As the method requires at least 4 points, there is a natural *restriction* N ⩾ 4.

$\left.\begin{array}{l} X \\ Y \end{array}\right\}$: [real one-dimensional arrays. Both arrays must have length at least N in the REAL declaration in your calling program]

Ref.
p. 19

X(I) and Y(I) are used to contain the i(th) data-point (x_i, y_i).

Thus, using the specimen problem as an example:

X(1) = 0.00, Y(1) = 4.0000
X(2) = 0.08, Y(2) = 3.9746

and so on.

Comments:
There are *restrictions* on the *x*-values.
   (i) they must all be distinct, and
   (ii) they must be arranged either in ascending order or in descending order.

i.e.   X(1) < X(2) < . . . < X(N)
or   X(1) > X(2) > . . . > X(N).

The error parameter
    IFAIL: [an integer variable]
       IFAIL is the error parameter described in §3.2. It is recommended that you set

       IFAIL = 0

       before you call D01GAF. Then in the event of the routine failing, your program will stop and print one of the following error messages:

Error message	Meaning	Advice
IFAIL = 1	N < 4.	Check your value of N, and that you have at least 4 data points.
IFAIL = 2	Your *x*-values are in the wrong order.	Check that your *x*-values are strictly increasing or decreasing.
IFAIL = 3	Two *x*-values are the same.	See restriction (i) on X, and check your *x*-values.

Parameters to be examined after calling D01GAF
    ANS: [a real variable]
       Assuming that everything has gone well, then ANS will contain the required approximation to the integral.

ER: [a real variable]
ER will contain an estimate of the likely absolute error in the approximation to the integral.

Specimen program

Program planning

1. *Declare* REAL X( ), Y( ), ANS, ER
   INTEGER N, IFAIL

2. *Read*   N, X, Y
   Set   IFAIL

3. *Call*   D01GAF

4. *Print*   ANS, ER

Comments:

(i) The following program will find an estimate for an integral where up to 100 data points can be specified. If you have more data points than this, then you will have to alter the REAL declaration at the beginning of the program accordingly.

(ii) You may need to make some changes to this program in order to make it run correctly on your computer. *See §3.4 for details.*

D01GAF specimen program

```
C       D01GAF: INTEGRAL FROM DATA POINTS

        REAL X(100), Y(100), ANS, ER
        INTEGER N, IFAIL, I

        WRITE (6,*) 'ENTER THE NUMBER OF DATA POINTS'
        READ (5,*) N
        WRITE (6,*)
     *   'ENTER THE (X,Y) VALUES, ONE PAIR PER LINE'
        DO 10 I = 1, N
          READ (5,*) X(I), Y(I)
     10 CONTINUE

        IFAIL = 0

        CALL D01GAF(X,Y,N,ANS,ER,IFAIL)

        WRITE (6,*) 'INTEGRAL = ', ANS
        WRITE (6,*) 'WITH AN ESTIMATED ERROR OF ', ER

        STOP

        END
```

D01GAF specimen run

```
ENTER THE NUMBER OF DATA POINTS
 9
ENTER THE (X,Y) VALUES, ONE PAIR PER LINE
 0.00   4.0000
 0.08   3.9746
 0.22   3.8153
 0.39   3.4719
 0.46   3.3014
 0.60   2.9412
 0.73   2.6094
 0.85   2.3222
 1.00   2.0000
INTEGRAL = 3.141593
WITH AN ESTIMATED ERROR OF 2.8194757E-05
```

Postscript

Accuracy:
You can see that in this example, the estimated error is less than 0.00003, so the answer is likely to be accurate to 4 decimal places; hence the answer 3.1416 could be quoted with some confidence.

Associated routines:
   (i) If you need a routine which finds the integral of a given function, then use D01AJF which is described in §6.2
   (ii) A multiple integral routine D01GBF is described in §10.5.
Ref. §8   (iii) If you often use integration routines, you would be advised to look at the appropriate *decision tree* in the *NAG Mini-Manual* first. This will help you choose the routine in the *NAG Manual* best suited to your problem.

6.4 Ordinary differential equations: D02BAF
[A medium routine]

The **purpose** of this routine is to find a numerical solution of a system of first-order ordinary differential equations.

$$\frac{dy_1}{dt} = f_1 (t, y_1, y_2, \ldots, y_n)$$

$$\frac{dy_2}{dt} = f_2 (t, y_1, y_2, \ldots, y_n) \tag{6.4}$$

$$\frac{dy_n}{dt} = f_n (t, y_1, y_2, \ldots, y_n)$$

for a chosen value of t, where the values of $y_1, y_2, \ldots, y_n$ are known for one particular initial value of t (say t_0).

Comment:
Problems where all the known information is given for one particular value of t are often referred to as *initial value problems*. This routine is restricted to problems of this kind.

Specimen problem

To find approximations to the values of y_1 and y_2 when $t = 10$, given the first-order differential equations

$$\frac{dy_1}{dt} = 2y_1(1 - y_2)$$

$$\frac{dy_2}{dt} = y_2(y_1 - 1)$$

with the initial conditions that $y_1 = 1$ and $y_2 = 3$ when $t = 0$.

Comment:
The right-hand sides of the equations defining dy_1/dt and dy_2/dt above could be functions of t, as well as y_1 and y_2, although the specimen problem does not demonstrate this point.

The **method** integrates the differential equations, using the Merson form of the Runge–Kutta method.

The **routine name** with parameters is
D02BAF (T, TEND, N, Y, TOL, FCN, W, IFAIL) .

Description of parameters

Parameters which require values before D02BAF is called

N: [an integer variable]
N is the number of differential equations.

In the specimen problem, there are 2 differential equations, so in this example N should be specified as 2.

T: [a real variable]
T should contain the value t_0 at which the initial conditions are given.

In the case of the specimen problem, the initial conditions are given at $t = 0$, so T should initially be specified as 0.0 for this problem.

Comment:
T has its initial value changed by the routine.

Ref.
p. 19

Y: [a real one-dimensional array. Its length must be at least N in the declaration in your calling program]

The array Y is used to contain the 'initial' values of $y_1, y_2, \ldots, y_n$ at $t = t_0$.

So, using the initial conditions of the specimen problem above,

$$Y(1) = 1.0 \quad \text{and} \quad Y(2) = 3.0.$$

TEND: [a real variable]

TEND should contain the value of t for which the values of $y_1, y_2, \ldots, y_n$ are required.

For instance, in the specimen problem, the values of y_1 and y_2 are required when $t = 10$, so TEND should be specified as 10.0.

Comment:

TEND can be less than the initial value t_0 if required.

TOL: [a real positive variable]

TOL should contain some small positive number. In general, the smaller the value that you give to TOL, the more accurate your result is likely to be.

Comment:

The connection between the value given to TOL and the accuracy of the answer is quite complicated. So it is suggested that if you want an answer which is likely to be accurate to k significant figures then run the program *at least twice* with TOL $= 10^{-k}$ and TOL $= 10^{-(k+1)}$ to see how many figures agree in the two answers. This is done in the specimen program.

[Further information about TOL can be found in the *NAG Manual*.]

Subroutines which require definition

Ref.
§2.2

FCN: [a subroutine. This must be declared as EXTERNAL at the beginning of your calling program]

The **purpose** of this subroutine is to specify the functions $f_1, f_2, \ldots, f_n$ of the differential equations given in (6.4).

The **name** of this routine with parameters is

FCN (T, Y, F)

The way to use this subroutine and its parameters and what these parameters mean, is best demonstrated by example. Suppose that you wanted to use this routine to specify the right-hand sides of the differential equations given in the specimen problem. Then a suitable subroutine FCN could be written as follows:

```
SUBROUTINE FCN (T, Y, F)
REAL T, Y(2), F(2)
F(1) = 2.0*Y(1)*(1.0 - Y(2))
F(2) = Y(2)*(Y(1) - 1.0)
RETURN
END
```

Comments on the subroutine FCN

(i) The length of the arrays Y and F in the REAL declaration *must* be the same as the number of differential equations to be solved.

Thus, for the specimen problem, where there are two differential equations, the subroutine FCN *must* have a declaration

REAL Y(2), F(2).

(ii) The statements 'F(1) = . . .' and 'F(2) = . . .' in the subroutine above specify the particular differential equations given in the specimen problem. In general, the array F(1), F(2), . . ., F(N) is used to specify the functions $f_1, f_2, \ldots, f_n$ of the differential equations given in (6.4).

(iii) You can see in the specimen subroutine FCN above, that no values have been assigned to T and Y. This is because the NAG routine D02BAF supplies values for these parameters when required. *In no circumstances should you attempt to assign values to T or Y in the subroutine FCN.*

Parameters associated with workspace

Ref.
§3.3

W: [a real two-dimensional array. It must have at least N rows and at least 7 columns in the REAL declaration in your calling program] This array is used as workspace.

So, for the specimen problem, where N = 2, a minimum declaration

REAL W(2,7)

would be needed in the program.

The error parameter

IFAIL: [an integer variable]
IFAIL is the error parameter described in §3.2. It is recommended that you set

IFAIL = 0

before you call D02BAF. Then in the event of the routine failing, your program will stop and print one of the following error messages:

Error message	Meaning	Advice
IFAIL = 1	*Either* TOL $\leq$ 0.0 *or* N $\leq$ 0.	Check your values of N and TOL.
IFAIL = 2 ⎫ IFAIL = 3 ⎭	The routine cannot do the problem with the value specified for TOL.	First try increasing TOL. If that doesn't work, get some help.
IFAIL = 4	A serious error has occurred in the routine.	Check all subroutine calls and dimensions of arrays. Get some help.

Parameters to be examined after calling D02BAF

Y: [the one-dimensional array described earlier]
Originally, this array was used to hold the initial y-values. After the routine has been called, if all has gone well, then the array Y(1), Y(2), . . ., Y(N) will contain the values of $y_1, y_2, . . ., y_n$ corresponding to $t =$ TEND.

Specimen program

Program planning

1. *Declare*   REAL T, TEND, Y( ), TOL, W(,7)
INTEGER N, IFAIL
EXTERNAL FCN

2. *Read*   N, T, TEND, Y, TOL
   Set   IFAIL

3. *Call*   D02BAF

4. *Print*   TEND, Y

5. *Write routine* FCN

Comments:
   (i) The following program can be used for a system of up to 5 differential equations. If solutions are required for larger systems, then you will have to alter the sizes of the arrays Y and W in the REAL declaration in the calling program accordingly.
   (ii) Each time you want to solve a new system of differential equations, you will have to change the subroutine FCN. In this case, check that in the REAL declaration the lengths of the arrays Y and F in FCN are the *same* as the number of differential equations to be solved.
   (iii) You may need to make some changes to this program in order to make it run correctly on your computer. *See §3.4 for details.*

D02BAF specimen program

```
C       D02BAF: ORDINARY DIFFERENTIAL EQUATIONS
C                  WITH INITIAL CONDITIONS

        REAL T, TEND, Y(5), TOL, W(5,7)
        INTEGER N, IFAIL, I
        EXTERNAL FCN

        WRITE (6,*)
    *   'ENTER THE NUMBER OF DIFFERENTIAL EQUATIONS'
        READ (5,*) N
        WRITE (6,*) 'ENTER THE INITIAL VALUE OF T'
        READ (5,*) T
        WRITE (6,*) 'ENTER THE INITIAL VALUES OF Y'
        READ (5,*) (Y(I),I=1,N)
        WRITE (6,*) 'ENTER THE FINAL VALUE OF T'
        READ (5,*) TEND
        WRITE (6,*) 'ENTER THE VALUE OF TOL'
        READ (5,*) TOL

        IFAIL = 0

        CALL D02BAF(T,TEND,N,Y,TOL,FCN,W,IFAIL)

        WRITE (6,*) 'THE Y VALUES CORRESPONDING TO T = ',
    *   TEND, ' ARE'
        WRITE (6,*) (Y(I),I=1,N)

        STOP

        END

        SUBROUTINE FCN(T,Y,F)
        REAL T, Y(2), F(2)
        F(1) = 2.0*Y(1)*(1.0-Y(2))
        F(2) = Y(2)*(Y(1)-1.0)
        RETURN
        END
```

D02BAF specimen runs

(a) with TOL = 10^{-4}

```
ENTER THE NUMBER OF DIFFERENTIAL EQUATIONS
 2
ENTER THE INITIAL VALUE OF T
 0.0
ENTER THE INITIAL VALUES OF Y
 1.0   3.0
ENTER THE FINAL VALUE OF T
 10.0
ENTER THE VALUE OF TOL
 1.0E-4
THE Y VALUES CORRESPONDING TO T = 10.00000 ARE
3.143382, 0.3485690
```

(b) with TOL $= 10^{-5}$

```
ENTER THE NUMBER OF DIFFERENTIAL EQUATIONS
 2
ENTER THE INITIAL VALUE OF T
 0.0
ENTER THE INITIAL VALUES OF Y
 1.0    3.0
ENTER THE FINAL VALUE OF T
 10.0
ENTER THE VALUE OF TOL
 1.0E-5
THE Y VALUES CORRESPONDING TO T = 10.00000 ARE
3.144279, 0.3488043
```

Postscript

Accuracy:

Looking at the two sets of results obtained with two different tolerances, the corresponding y-values agree to 3 significant figures. So, in this example, you could be fairly confident that $y_1 = 3.14$ and $y_2 = 0.349$ when $t = 10$.

Comment:

D02BAF is, in fact, much more flexible than it appears on first reading. This routine (and most of the other differential equation routines in the *NAG Manual*) solves systems of first-order equations. The reason for this is that a differential equation of order n can usually be reformulated as a system of n first-order differential equations.

Take, for instance, the third-order differential equation.

$$\frac{d^3z}{dt^3} + z\frac{d^2z}{dt^2} + 5\left(1 - \left(\frac{dz}{dt}\right)^2\right) = 0 .$$

If you put

$$y_1 = z \qquad\qquad \text{(i)}$$

$$y_2 = \frac{dz}{dt} \qquad\qquad \text{(ii)}$$

$$y_3 = \frac{d^2z}{dt^2} \qquad\qquad \text{(iii)} ,$$

then the equation above can be rewritten

$$\frac{dy_3}{dt} = -y_1 y_3 - 5\left(1 - y_2^2\right) ,$$

and (ii) and (iii) can be rewritten

$$\frac{dy_1}{dt} = y_2$$

and $\quad \dfrac{\mathrm{d}y_2}{\mathrm{d}t} = y_3$.

Thus, the original third-order differential equation can be reformulated as a system of 3 first-order equations which can be 'solved' using D02BAF.

More generally, if you have an n(th) order differential equation to solve, then use the technique described above to reformulate the problem as a system of n first-order equations. Then, use D02BAF to solve the reformulated problem.

It goes almost without saying that a single first-order differential equation is a system consisting of one equation.

Associated routines:

(i) If you need the values of y corresponding to several different values of t, then use D02BBF, which is described in the Addendum on the next page.

(ii) If you suspect that your equations are stiff (i.e. that the solution has a transient part which dies away rapidly), then use D02EAF. You will find details of this routine in the *NAG Manual*.

(iii) If you have a boundary value problem then try using D02GAF. You can find details of this routine in the *NAG Manual*.

Addendum: D02BBF
[A difficult routine]

D02BAF is explicitly designed to find y-values corresponding to *one particular* value of t. Although it would be possible to adapt the specimen program given for D02BAF to find y-values corresponding to *several different* values of t, you are strongly advised, in these circumstances, to use D02BBF instead. This routine is designed to do this particular job, and does it more efficiently than D02BAF.

As most of the parameter descriptions in D02BBF are exactly the same as those in D02BAF, it is not worth describing them all again. The extra information needed to use D02BBF is given below.

The **purpose** of this routine is to find a numerical solution to the first-order ordinary differential equations given in equation (6.4) for *several different* values of t, where initial conditions are given.

Specimen problem

To find an approximation to y_1 and y_2 *when $t = 0, 1, \ldots, 10$*, given the first-order differential equations.

$$\frac{dy_1}{dt} = 2y_1(1 - y_1)$$

$$\frac{dy_2}{dt} = y_2(y_1 - 1)$$

with the initial conditions $y_1 = 1$ and $y_2 = 3$ when $t = 0$.

The **routine name** with parameters is

D02BBF (T, TEND, N, Y, TOL, IR, FCN, OUTPUT, W, IFAIL).

Parameters

Comment:

Consult D02BAF for the descriptions of the parameters T, N, Y, TOL, FCN, and W. An amended description for TEND and the meaning of the different IFAIL values is given below.

TEND: [a real variable]

TEND should contain the *final* value of t for which the y-values are required.

For instance, in the specimen problem, the final value of t for which y_1 and y_2 are required is when $t = 10$. So TEND should be specified as 10.0.

Comment:

TEND can be less than the initial value t_0 if required.

IFAIL: [an integer variable]
The different error messages which can occur are given below.

Error message	Meaning	Advice
IFAIL = 1	*Either* TOL $\leq$ 0 *or* N $\leq$ 0 *or* IR $\neq$ 0, 1, *or* 2.	Check your values of TOL, N and IR.
IFAIL = 2 ⎱ IFAIL = 3 ⎰	The routine cannot do the problem with the value specified for TOL.	Try increasing TOL. If this doesn't work, get some help.
IFAIL = 4	T and TEND have the same value initially.	Check the programming of the subroutine OUTPUT.
IFAIL = 5	The value of TSOL is moving away from TEND.	Check the programming of the subroutine OUTPUT.
IFAIL = 6	A serious error has occurred in the routine.	Check all subroutine calls and dimensions of arrays. Get some help.

Other parameters needed for D02BBF

IR: [an integer variable]
IR must have the value of 0, 1, or 2, according to the kind of control you wish to exercise over the error in the answers. It is recommended that you set

IR = 0 .

**Ref.
§5.4(b)** In this event, if all the *y*-values are numerically less than 1.0, then TOL is used to control the absolute error in the final answer. Otherwise, TOL is used to control the relative accuracy in the answer.

For further details of this parameter, consult the *NAG Manual*.

OUTPUT: [a subroutine. This must be declared as EXTERNAL at the
**Ref.
§ 2.2** beginning of your calling program]

The **purpose** of OUTPUT is
(i) to specify the values of *t* at which you require corresponding *y*-values, and
(ii) to get the *y*-values printed (or plotted, if required).

The **name** of this subroutine with parameters is

OUTPUT (TSOL, Y)

where TSOL contains the current value of *t* at which the *y*-values are required, and the array Y(1), Y(2), . . ., Y(N) contains the corresponding *y*-values.

To understand how to write a suitable subroutine OUTPUT, first it is necessary to understand roughly the way in which D02BBF works.

(i) Initially, D02BBF sets TSOL = T (the initial value of t) before it calls OUTPUT.

(ii) The following process is then repeated while TSOL $\leqslant$ TEND:
  (a) OUTPUT
* gets the y-values printed for the current value of TSOL, and
* sets TSOL to the next t-value for which you require a solution.
  (b) D02BBF now calculates new y-values corresponding to TSOL, and discards the old ones.

Your job is to write a suitable subroutine OUTPUT.

In the case of the specimen problem, the y-values are required for $t = 0$, 1, . . ., 10. It would be very convenient to have a variable STEP in the subroutine OUTPUT, so that the TSOL values could be determined by the statement

   TSOL = TSOL + STEP ,

where STEP would have the value 1.0 in this instance. It would also be convenient to have the value of TEND in the subroutine to determine when to stop the printing. As neither STEP nor TEND are parameters of the subroutine, these values would have to be carried by using the COMMON statement

Ref.
§2.5

      COMMON STEP, TEND

in both the subroutine and in the calling program.

Using this facility, a suitable subroutine OUTPUT could be written as follows:

```
SUBROUTINE OUTPUT(TSOL,Y)
REAL TSOL, Y(2), STEP, TEND
INTEGER I
COMMON STEP, TEND
WRITE (6,'(1X,6E13.4)') TSOL, (Y(I),I=1,2)
TSOL = TSOL + STEP
IF (ABS(TSOL-TEND).LE.1.0E-4) THEN
   TSOL = TEND
END IF
RETURN
END
```

Note: Due to the way in which D02BBF is constructed, the results have to be printed in the subroutine OUTPUT, as after a call to OUTPUT, D02BBF destroys the current y-values.

Comments on the subroutine OUTPUT

(i) The length of the array Y in the declaration *must* be the same as the number of differential equations to be solved.

Thus, for the specimen problem, where there are two differential equations, the subroutine OUTPUT *must* have a declaration

REAL Y(2) .

(ii) Note that the NAG routine D02BBF supplies the required values for Y. *So in no circumstances should you attempt to assign values to Y in the subroutine OUTPUT.*

(iii) If you set TSOL to a value greater than TEND, then OUTPUT will not be called by D02BBF again. So, if you want the *y*-values printed for T = TEND, then you must arrange for TSOL to be set exactly to TEND in the subroutine OUTPUT. The two lines

IF(ABS(TSOL−TEND).LE.1.0E−4) THEN
TOL=TEND

in the subroutine above, ensure that the *y*-values corresponding to T=TEND are printed.

Specimen program

Program planning

1. *Declare*	REAL T, TEND, Y( ), TOL, W(,7)
	INTEGER N, IR, IFAIL
	EXTERNAL FCN, OUTPUT
2. *Read*	N, T, TEND, Y, TOL
Set	IFAIL, IR
3. *Call*	D02BBF
4. *Write routine* FCN	
Write routine OUTPUT	

Comments:

(i) The following program can be used for a system of up to 5 differential equations. If solutions are required for larger systems, you will have to alter the sizes of the arrays Y and W in the REAL declaration in the calling program accordingly.

(ii) Each time you want to solve a new system of differential equations you will have to change the subroutines FCN and OUTPUT. In this case, check that the sizes of the arrays Y and F in FCN, and of the array Y in OUTPUT are the same as the number of differential equations which you are solving.

(iii) You may need to make some changes to this program in order to make it run correctly on your computer. *See §3.4 for details*

D02BBF specimen program

```
C      D02BBF: ORDINARY DIFFERENTIAL EQUATIONS
C                WITH INITIAL CONDITIONS
       REAL T, TEND, Y(5), TOL, W(5,7), STEP
       INTEGER N, IR, IFAIL, I
       EXTERNAL FCN, OUTPUT
       COMMON STEP, TEND

       IR = 0

       WRITE (6,*)
     * 'ENTER THE NUMBER OF DIFFERENTIAL EQUATIONS'
       READ (5,*) N
       WRITE (6,*) 'ENTER THE INITIAL VALUE OF T'
       READ (5,*) T
       WRITE (6,*) 'ENTER THE INITIAL VALUES OF Y'
       READ (5,*) (Y(I),I=1,N)
       WRITE (6,*) 'ENTER THE FINAL VALUE OF T'
       READ (5,*) TEND
       WRITE (6,*) 'ENTER THE STEP SIZE REQUIRED FOR T'
       READ (5,*) STEP
       WRITE (6,*) 'ENTER THE VALUE OF TOL'
       READ (5,*) TOL

       IFAIL = 0

       WRITE (6,*) '    T              Y VALUES'

       CALL D02BBF(T,TEND,N,Y,TOL,IR,FCN,OUTPUT,W,IFAIL)

       STOP

       END

       SUBROUTINE FCN(T,Y,F)
       REAL T, Y(2), F(2)
       F(1) = 2.0*Y(1)*(1.0-Y(2))
       F(2) = Y(2)*(Y(1)-1.0)
       RETURN
       END

       SUBROUTINE OUTPUT(TSOL,Y)
       REAL TSOL, Y(2), STEP, TEND
       INTEGER I
       COMMON STEP, TEND
       WRITE (6,'(1X,6E13.4)') TSOL, (Y(I),I=1,2)
       TSOL = TSOL + STEP
       IF (ABS(TSOL-TEND).LE.1.0E-4) THEN
         TSOL = TEND
       END IF
       RETURN
       END
```

D02BBF specimen runs

You are recommended (as in D02BAF) to run your program at least twice with different tolerances to see how many significant figures agree in the two answers.

(a) with TOL = 10^{-4}

```
ENTER THE NUMBER OF DIFFERENTIAL EQUATIONS
 2
ENTER THE INITIAL VALUE OF T
 0.0
ENTER THE INITIAL VALUES OF Y
 1.0  3.0
ENTER THE FINAL VALUE OF T
 10.0
ENTER THE STEP SIZE REQUIRED FOR T
 1.0
ENTER THE VALUE OF TOL
 1.0E-4
T            Y VALUES
0.0000E+00   0.1000E+01   0.3000E+01
0.1000E+01   0.7738E-01   0.1464E+01
0.2000E+01   0.8501E-01   0.5780E+00
0.3000E+01   0.2909E+00   0.2493E+00
0.4000E+01   0.1446E+01   0.1872E+00
0.5000E+01   0.4052E+01   0.1439E+01
0.6000E+01   0.1757E+00   0.2259E+01
0.7000E+01   0.6532E-01   0.9089E+00
0.8000E+01   0.1472E+00   0.3667E+00
0.9000E+01   0.6503E+00   0.1876E+00
0.1000E+02   0.3143E+01   0.3486E+00
```

(b) with TOL = 10^{-5}

```
ENTER THE NUMBER OF DIFFERENTIAL EQUATIONS
 2
ENTER THE INITIAL VALUE OF T
 0.0
ENTER THE INITIAL VALUES OF Y
 1.0  3.0
ENTER THE FINAL VALUE OF T
 10.0
ENTER THE STEP SIZE REQUIRED FOR T
 1.0
ENTER THE VALUE OF TOL
 1.0E-5
T            Y VALUES
0.0000E+00   0.1000E+01   0.3000E+01
0.1000E+01   0.7735E-01   0.1464E+01
0.2000E+01   0.8498E-01   0.5780E+00
0.3000E+01   0.2909E+00   0.2493E+00
0.4000E+01   0.1447E+01   0.1872E+00
0.5000E+01   0.4051E+01   0.1439E+01
0.6000E+01   0.1756E+00   0.2259E+01
0.7000E+01   0.6531E-01   0.9088E+00
0.8000E+01   0.1472E+00   0.3667E+00
0.9000E+01   0.6506E+00   0.1876E+00
0.1000E+02   0.3144E+01   0.3488E+00
```

Comment:
Looking at the two sets of results above, as the corresponding *y*-values
agree to 3 significant figures, then it should be reasonably safe to quote
answers to 3 significant figures.

6.5 Curve-fitting, polynomial approximation: E02ADF [A medium routine]

The **purpose** of this routine is to attempt to find a suitable polynomial
approximation to a given set of data points $(x_1, y_1), (x_2, y_2), \ldots, (x_n, y_n)$.

Comment:
This routine, instead of finding the coefficients $c_0, c_1, \ldots$ of

$$y = c_0 + c_1 x + c_2 x^2 + \ldots,$$

finds the coefficients of the equivalent normalized Chebyshev series.
How to choose the best polynomial for your data, and what to do with the
Chebyshev coefficients is discussed along with the specimen program.

Specimen problem

To find a suitable polynomial fit to the 9 data points $(-4.2, 43), (-2.8, 24), (-2.0, 9.2), (-1.3, 2.1), (0.0, 1.1), (0.8, 6.1), (2.2, 16), (3.0, 32)$ and
$(4.1, 58)$.

The **method** used is basically a least-squares method.

The **routine name** with parameters is

    E02ADF (N, KPLUS1, NROWS, X, Y, W, WORK1, WORK2, A, S,
        IFAIL)

Description of parameters

Parameters which require values before E02ADF is called

  N: [an integer variable]
    N is the number of data points.

    In the specimen problem, there are 9 data points, so N should be given the
    value 9.

  X⎫ [real one-dimensional arrays. Both arrays must have length at least N in
  Y⎭ the REAL declaration in your calling program]
Ref.   X(I) and Y(I) are used to contain the *i*(th) data point (x_i, y_i).
p. 19

continued

Thus, using the specimen problem as an example

$$X(1) = -4.2, \quad Y(1) = 43.0$$
$$X(2) = -2.8, \quad Y(2) = 24.0$$

and so on.

Comment:
The points must be read in so that the *x*-values are non-decreasing, i.e.
$x_1 \leqslant x_2 \leqslant x_3 \leqslant \ldots \leqslant x_n$.

Also x_1 must be less than x_n.

W: [a positive real one-dimensional array. Its length must be at least N in the declaration in your calling program]
It is possible when using this routine to weight points so that some are more important than others. W is an array which holds these weighting factors for each data point. So W(I) will hold the weighting factor for the data point (x_i, y_i).

In the case of the specimen problem, all the points are equally weighted, so W(1) = 1.0, W(2) = 1.0, . . ., W(9) = 1.0.

Comment:
All the weights must be positive.

KPLUS1: [an integer variable]
Suppose that you wanted to look at polynomials up to degree K which fitted your data. Then KPLUS1 should be given the value (K + 1).

In practical terms, the routine is worth using only if you are looking for a simple polynomial fit to your data – say up to degree 6. So, you are recommended to give KPLUS1 a value of 7 or less.

Comment:
(i) KPLUS1 > 0.
(ii) KPLUS 1 must be less than the number of distinct X values in the data.

NROWS: [an integer variable]
NROWS must specify the size of the first dimension of the array A as declared in your calling program. The array A is described later.

Ref.
§2.4

Suppose, for example, your calling program had a declaration

REAL A(8, 7).

In this case, a statement

NROWS = 8

would also be needed in your program.

Parameters associated with workspace

WORK1: [a real two-dimensional array. It must have at least 3 rows and at least N columns in the REAL declaration in your calling program]

WORK2: [a real two-dimensional array. It must have at least 2 rows and at least (K + 1) columns in the REAL declaration in your calling program]

Ref.
§.3.3
Both these arrays are used as workspace.

Comment:

A declaration

REAL WORK1 (3, 100), WORK2 (2, 7)

will cope with any problem which has up to 100 data points, and a polynomial fit of degree less than 7.

The error parameter

IFAIL: [an integer variable]

IFAIL is the error parameter described in §3.2. It is recommended that you set

IFAIL = 0

before you call E02ADF. Then in the event of the routine failing, your program will stop and print one of the following error messages:

Error message	Meaning	Advice
IFAIL = 1	Not all the weights W(1), . . ., W(N) are positive.	Check that your values of W are all positive. [See restriction on W.]
IFAIL = 2	The values X(1), X(2), . . ., X(N) are arranged in the wrong order.	Rearrange the data points so that X-values are in non-decreasing order. [See restriction on X.]
IFAIL = 3	You have specified all the X values the same.	The routine cannot cope with this. In this event you could write down the equation yourself.
IFAIL = 4	*Either* KPLUS1 < 1 *or* KPLUS1 is larger than the number of distinct X values.	Check your value for KPLUS1.

Table—*continued*

Error message	Meaning	Advice
IFAIL = 5	The value of NROWS is less than KPLUS1.	*Either* increase the size of A in the declaration, and the corresponding value of NROWS *or* decrease the value of KPLUS1 to satisfy the restriction.

Parameters to be examined after calling E02ADF

A: [a real two-dimensional array. It must have at least $(K + 1)$ rows and at least $(K + 1)$ columns in the REAL declaration in your calling program] If all has gone well in the routine, then the $(j + 1)$(th) row of A will hold the coefficients of a polynomial of degree j. A description of the exact form of this polynomial, and how to convert it into something recognizable, follows after the program.

Let it suffice to say for the moment that the coefficients of the polynomial of zero degree will be held in $A(1,1)$, the coefficients of the first degree polynomial will be held in $A(2, 1)$ and $A(2, 2)$, the coefficients of the second degree polynomial in $A(3, 1)$, $A(3, 2)$ and $A(3, 3)$, and so on.

You can get the coefficients of all the polynomials up to your chosen value of K if you print the coefficients

a_{11}
$a_{21}\ a_{22}$
$a_{31}\ a_{32}\ a_{33}$
$a_{41}\ a_{42}\ a_{43}\ a_{44}$
$\vdots$
$a_{k+1,1}\ a_{k+1,2}\ \cdots\ a_{k+1,\ k+1}$

S: [a real one-dimensional array. Its length must be at least $(K + 1)$ in the declaration in your calling program] The contents of $S(I + 1)$ give you a measure of the y-deviation of the points from the polynomial of degree I. In particular, $S(1)$ will give a measure of the error in using the polynomial of degree zero as an approximation of your points, $S(2)$ will give a measure of the error in using the first degree polynomial, $S(3)$ will give a measure of the error in using the second degree polynomial, and so on.

Comment:

When running your program, you can compare the values in $S(1)$, $S(2)$, $S(3)$ . . . to see which polynomial best fits your data. This is discussed more in the postscript at the end of the specimen program.

Specimen program

Program planning

1. *Declare* REAL X( ), Y( ), W( ), WORK1(3,), WORK2(2,),
          A(,), S( )
          INTEGER N, KPLUS1, NROWS, IFAIL, K

2. *Read*   N, X, Y, W, K
   Set   KPLUS1, NROWS, IFAIL

3. *Call*   E02ADF

4. *Print* A, S

Comments:

(i) The following program can be used to obtain the coefficients of polynomial approximations of degree less than 7, where up to 200 data points may be specified. If you have more than 200 data points, then you should change the size of the arrays X, Y, W and WORK1 in the REAL declaration in the program accordingly.

(ii) If your polynomial is likely to have degree greater than 6, you should consider whether a routine using spline functions is not preferable. [See the notes at the end of the program.]

(iii) You may need to make some changes to this program in order to make it run correctly on your computer. *See §3.4 for details.*

E02ADF specimen program

```
C       E02ADF: FITTING A POLYNOMIAL

        REAL X(200), Y(200), W(200), WORK1(3,200), WORK2(2,7),
     *      A(7,7), S(7)
        INTEGER N, KPLUS1, NROWS, IFAIL, I, J, K, IP

        NROWS = 7

        WRITE (6,*) 'ENTER NUMBER OF POINTS'
        READ (5,*) N
        WRITE (6,*) 'ENTER THE (X,Y) VALUES AND THEIR WEIGHTS'
        DO 10 I = 1, N
          READ (5,*) X(I), Y(I), W(I)
     10 CONTINUE
        WRITE (6,*)
     *   'ENTER THE HIGHEST DEGREE POLYNOMIAL REQUIRED'
        READ (5,*) K
        KPLUS1 = K + 1

        IFAIL = 0
```

continued

```
              CALL E02ADF(N,KPLUS1,NROWS,X,Y,W,WORK1,WORK2,A,S,
         *            IFAIL)

         DO 20 I = 1, K + 1
            IP = I - 1
            WRITE (6,*)
            WRITE (6,*)
         *    'CHEBYSHEV COEFFS FOR POLYNOMIAL OF DEGREE', IP
            WRITE (6,*) (A(I,J),J=1,I)
            WRITE (6,*) 'WITH RMS RESIDUAL ', S(I)
      20 CONTINUE

         STOP

         END
```

E02ADF specimen run

```
ENTER NUMBER OF POINTS
 9
ENTER THE (X,Y) VALUES AND THEIR WEIGHTS
 -4.2  43.0  1.0
 -2.8  24.0  1.0
 -2.0   9.2  1.0
 -1.3   2.1  1.0
  0.0   1.1  1.0
  0.8   6.1  1.0
  2.2  16.0  1.0
  3.0  32.0  1.0
  4.1  58.0  1.0
ENTER THE HIGHEST DEGREE POLYNOMIAL REQUIRED
 4

CHEBYSHEV COEFFS FOR POLYNOMIAL OF DEGREE 0
42.55556
WITH RMS RESIDUAL 19.75404

CHEBYSHEV COEFFS FOR POLYNOMIAL OF DEGREE 1
42.45842, 7.256409
WITH RMS RESIDUAL 20.46190

CHEBYSHEV COEFFS FOR POLYNOMIAL OF DEGREE 2
52.14209, 5.627827, 24.89593
WITH RMS RESIDUAL 2.715458

CHEBYSHEV COEFFS FOR POLYNOMIAL OF DEGREE 3
52.05722, 5.754294, 24.97493, 1.310750
WITH RMS RESIDUAL 2.610052

CHEBYSHEV COEFFS FOR POLYNOMIAL OF DEGREE 4
52.20231, 5.718202, 25.06160, 1.303564, -0.4853010
WITH RMS RESIDUAL 2.859159
```

Postscript

Comment:

(i) *Choosing your polynomial*

Examination of the S-values (the root mean square residuals) will help you to decide which polynomial is best suited to your data.

For example, in the specimen run above, S drops from 20.5 for a polynomial of degree 1, to 2.7 for a polynomial of degree 2, and then does not *substantially* decrease. So, in this case, it is reasonable to argue that a quadratic best fits your data.

(ii) *What to do next*

When you have chosen the degree of polynomial which best fits your data, you will have to decide whether

(a) you need an equation of the form

$$y = c_0 + c_1x + c_2x^2 + \dots,$$

or

(b) you just want a point (or points) on the curve of best fit

or

(c) you want a graph of this curve.

There are NAG routines to find points on the curve or a graph of the curve. However, if you *really* need an equation, then you are going to have to do some hard work.

The routine gives you the coefficients $a_1, a_2, a_3 \dots$ of the normalized Chebyshev series

$$f(x) = \tfrac{1}{2} a_1 T_0(\bar{x}) + a_2 T_1(\bar{x}) + a_3 T_2(\bar{x}) + \dots \tag{6.5}$$

In the formula above,

(i) $\bar{x} = \dfrac{2x - x_{max} - x_{min}}{x_{max} - x_{min}},$

where x_{max} and x_{min} are the maximum and minimum values of your data, and

(ii) $T_1, T_2, \dots$ are Chebyshev polynomials

$$T_k(x) = \cos k\theta \quad (k = 0, 1, 2, \dots) \text{ where } x = \cos \theta.$$

In particular, the Chebyshev polynomials which you might need are

$$T_0(x) = 1$$
$$T_1(x) = x$$
$$T_2(x) = 2x^2 - 1$$
$$T_3(x) = 4x^3 - 3x$$
$$T_4(x) = 8x^4 - 8x^2 + 1$$

$$T_5(x) = 16x^5 - 20x^3 + 5x$$
$$T_6(x) = 32x^6 - 48x^4 + 18x^2 - 1 \ .$$

In the specimen problem, where $a_1 \simeq 52.14$, $a_2 \simeq 5.63$, $a_3 \simeq 24.90$, and x_{max} = 4.1, x_{min} = −4.2.

then $$\bar{x} = \frac{2x - (4.1) - (-4.2)}{4.1 - (-4.2)}$$

$$= (2x + 0.1)/8.3 \ .$$

So, the required second-degree polynomial equation is

$$y = \tfrac{1}{2}a_1 T_0(\bar{x}) + a_2 T_1(\bar{x}) + a_3 T_2(\bar{x})$$

$$= \tfrac{1}{2}(52.14)(1) + 5.63(\bar{x}) + 24.90 \times (2\bar{x}^2 - 1)$$

$$= \tfrac{1}{2} (52.14) + 5.63 \times \left(\frac{2x + 0.1}{8.3}\right) + 24.90 \times \left(2 \times \left(\frac{2x + 0.1}{8.3}\right)^2 - 1\right)$$

$$\simeq 1.25 + 1.65x + 2.89x^2.$$

It should be clear from this example, that if you only want to get points on a graph from this formula, then it would be easier to apply the appropriate NAG routine directly to the data $a_1, a_2, a_3 \ . \ . \ .$ to do this job.

(iii) Evaluation of a polynomial $\Sigma c_r x^r$ is usually more accurate if this polynomial is evaluated in its Chebyshev series form $\Sigma a_j T_j(\bar{x})$. It is for this reason, that E02ADF computes the coefficients of the Chebyshev series rather than those for the more obvious power series.

Associated routines:

(i) You might well require a single point or a set of points which lie on the curve of best fit, rather than the equation of the curve. In this event, use E02AEF which follows in §6.6

(ii) If you want a graph of the curve of best fit, then use E02AEF (described in §6.6) to get a set of points on the curve. Then use J06BAF and J06CAF to plot the graph. These are described in Chapter 7.

(iii) Often, a good low-degree polynomial approximation cannot be found. In this event, you should try using a routine which fits spline functions. The recommended routine is E02BAF, which can be found in the *NAG Manual*.

(iv) If you suspect that the function underlying your data is not a polynomial but some other function of x which is linear in the coefficients a, b, and c (e.g. $y = a + bx + c/x$), then use G02CJF which is described in §6.13.

(v) If you suspect that the function underlying your data is not polynomial, and is non-linear in the coefficients a, b, and c (e.g. $y = a + b/(x + c)$), then use E04FDF, which is described in §10.7.

6.6 Points from Chebyshev coefficients: E02AEF
[An easy routine]

The **purpose** of this routine is to evaluate

$$y = \tfrac{1}{2}a_1 T_0(\bar{x}) + a_2 T_1(\bar{x}) + \ldots a_{n+1} T_n(\bar{x})$$

where $a_1, a_2, \ldots, a_n$ are known, and $T_k(\bar{x})$ ($k = 0, 1, 2 \ldots$) are the Chebyshev polynomials described in E02ADF (§6.5).

Comment:
This routine essentially finishes off the job started in the previous routine E02ADF.

(i) It assumes that you have chosen a suitable n(th) degree polynomial of best fit using E02ADF. Hence, you know the coefficients $a_{n+1,1}, a_{n+1,2}, \ldots, a_{n+1,n+1}$ of the Chebyshev form of this polynomial.

(ii) It also assumes that you wish to evaluate this polynomial for given values of x. This routine *restricts* x-values to lie in the interval $[x_{min}, x_{max}]$, where x_{min} and x_{max} are the minimum and maximum x-values in the original data which you supplied to E02ADF.

Specimen problem

The specimen program in E02ADF produced the 'Chebyshev coefficients'

$$a_{31} = 52.14209 \quad a_{32} = 5.627827 \quad a_{33} = 24.89593$$

for a polynomial degree of 2.

To find y-values on the curve

$$y = \tfrac{1}{2} \times 52.14209 T_0(\bar{x}) + 5.627827 T_1(\bar{x}) + 24.89593 T_2(\bar{x})$$

corresponding to $x = -4.1, -3.0, -1.0,$ and 2.7.

Comments:
(i) In E02ADF, the coefficients of the polynomial of degree 2 were located in a_{31}, a_{32}, a_{33}. However, for the purpsoe of E02AEF, this information must be put in a_1, a_2, a_3.

(ii) The original x-values specified in the specimen problem in E02ADF all lay in the interval $[-4.2, 4.1]$. So, for the specimen problem, the x-values for which y is required all have to lie in this interval.

The **method** used to evaluate the polynomial is based on a three-term recurrence relation due to Clenshaw.

The **routine name** with parameters is

E02AEF (NPLUS1, A, XBAR, YPOL, IFAIL).

Description of parameters

Parameters which require values before E02AEF is called

NPLUS1: [an integer variable]
Suppose the polynomial of best fit has degree N. Then NPLUS1 must be given the value (N + 1).

In the specimen problem, the polynomial degree 2, so NPLUS1 = 3.

A: [a real one-dimensional array. Its length must be at least (N + 1) in the REAL declaration in your calling program]

Ref.
p. 19
The array A is used to contain the 'Chebyshev coefficients' found by E02ADF.

In the specimen problem, these coefficients are 52.14209, 5.627827 and 24.89593, so A(1) = 52.14209, A(2) = 5.627827, and A(3) = 24.89593.

Comment:
If E02ADF and E02AEF are both called in the same program, then the 'Chebyshev coefficients' which are returned to a two-dimensional array by E02ADF, *must* be transferred into a one-dimensional array before E02AEF is called. To see how this is done, see the specimen program for E02ADF/E02AEF at the end of this description.

XBAR: [a real variable]
XBAR must contain the normalized value

$$\text{XBAR} = \frac{2x - x_{max} - x_{min}}{x_{max} - x_{min}}$$

of any value x at which a Chebyshev polynomial is to be evaluated, where x_{max} and x_{min} are the maximum and minimum x-values in the original data.

Note: XBAR must have a value between −1 and 1.

Comment:
This means that you are going to have to do some work before the routine is called. You will have to specify x_{max} and x_{min}, and the value of x whose y-value is required, before XBAR can be given a value.

> To minimize the error in calculating XBAR, you are asked to specify
>
> XBAR = ((X − XMIN) − (XMAX − X))/(XMAX − XMIN)
>
> (6.6)
>
> in your calling program, rather than the form given above.

In the specimen problem, XMIN $= -4.2$ and XMAX $= 4.1$ – the minimum and maximum values in the original data given on p. 69.

The error parameter

IFAIL: [an integer variable]

IFAIL is the error parameter described in §3.2. It is recommended that you set

IFAIL $= 0$

before you call E02AEF. Then in the event of the routine failing, your program will stop and print one of the following error messages:

Error message	Meaning	Advice
IFAIL = 1	XBAR is outside the range $[-1,1]$.	Check (a) you have specified XBAR using the form (6.6) and (b) that the value of X which you have specified lies between XMIN and XMAX.
IFAIL = 2	NPLUS1 < 1.	Check that N > 0 and NPLUS1 = N + 1.

Parameter to be examined after calling E02AEF

YPOL: [a real variable]

If all has gone well in the routine, then YPOL will contain the value of the approximating polynomial corresponding to the value you gave X (or XBAR).

Specimen program

Program planning

1. *Declare* REAL A( ), XBAR, YPOL
   INTEGER NPLUS1, IFAIL

2. *Read* N (degree of polynomial)
   Set NPLUS1
   Read A, XMIN, XMAX, M (number of points)
   X
   Set XBAR
   IFAIL

3. *Call* E02AEF
   Print YPOL

Comments:

(i) The following program will find the values of a polynomial up to degree 6 for different values of *x* in the range (XMIN, XMAX). If you need to evaluate a polynomial of higher degree (which is not advised), then you will have to change the size of the array A in the REAL declaration in the program accordingly.

(ii) You may need to make some changes to this program in order to make it run correctly on your computer. *See §3.4 for details.*

E02AEF specimen program

```
C       E02AEF: VALUES OF A POLYNOMIAL GIVEN
C               IN CHEBYSHEV FORM

        REAL A(7), XBAR, YPOL, XMIN, XMAX
        INTEGER NPLUS1, IFAIL, N, M, I

        WRITE (6,*) 'ENTER THE DEGREE OF THE POLYNOMIAL'
        READ (5,*) N
        NPLUS1 = N + 1
        WRITE (6,*) 'ENTER THE ', NPLUS1,
     *   ' CHEBYSHEV COEFFICIENTS'
        READ (5,*) (A(I),I=1,N+1)
        WRITE (6,*)
     *       'ENTER THE SMALLEST AND LARGEST X VALUES IN THE ORIGINAL DATA'
        READ (5,*) XMIN, XMAX
        WRITE (6,*) 'ENTER THE NUMBER OF Y VALUES REQUIRED'
        READ (5,*) M
        DO 10 I = 1, M
          WRITE (6,*) 'ENTER A VALUE FOR X'
          READ (5,*) X
          XBAR = ((X-XMIN)-(XMAX-X))/(XMAX-XMIN)

          IFAIL = 0

          CALL E02AEF(NPLUS1,A,XBAR,YPOL,IFAIL)

          WRITE (6,*) 'AT X = ', X,
     *       ', THE POLYNOMIAL GIVES Y = ', YPOL
   10   CONTINUE

        STOP

        END
```

E02AEF specimen run

```
ENTER THE DEGREE OF THE POLYNOMIAL
 2
ENTER THE 3 CHEBYSHEV COEFFICIENTS
 52.14209   5.627827    24.89593
ENTER THE SMALLEST AND LARGEST X VALUES IN THE ORIGINAL DATA
 -4.2    4.1
ENTER THE NUMBER OF Y VALUES REQUIRED
 4
ENTER A VALUE FOR X
 -4.1
```

```
AT X = -4.100000, THE POLYNOMIAL GIVES Y = 43.10406
ENTER A VALUE FOR X
-3.0
AT X = -3.000000, THE POLYNOMIAL GIVES Y = 22.33435
ENTER A VALUE FOR X
-1.0
AT X = -1.000000, THE POLYNOMIAL GIVES Y = 2.496029
ENTER A VALUE FOR X
2.7
AT X = 2.700000, THE POLYNOMIAL GIVES Y = 26.76829
```

Postscript

Associated routines:

(i) This program can be combined with the specimen program for E02ADF. This will enable you to

(a) read in your data points.

(b) obtain the coefficients of the Chebyshev series given in equation (6.5),

(c) choose which polynomial will best fit your data, and

(d) get a print out of points on the polynomial of best fit corresponding to any x-value in the data range $[x_{max}, x_{min}]$.

A program combining the use of E02ADF and E02AEF can be found below.

(ii) These routines can be followed by use of the graph plotting routines in the *NAG Graphical Supplement*. You can use J06BAF to plot the points, and J06CAF to put a curve through the points. You will find details of these routines in Chapter 7.

(iii) Another routine which does the same job as E02AEF is E02AKF, which is described in the *NAG Manual*. The reason that this routine has not been included here is because it requires knowledge of how Fortran arrays are stored, which is not covered in this book.

E02ADF/E02AEF specimen program

The following program

Ref.
§5.2

(a) reads a set of data points, their weights, and the highest degree polynomial required from a data file.

(b) asks you to choose the polynomial of best fit, and

(c) prints y-values on your chosen polynomial corresponding to x-values, which are supplied at run-time.

```
C     E02ADF/AEF: POINTS ON POLYNOMIAL OF BEST FIT

      REAL X(200), Y(200), W(200), WORK1(3,200), WORK2(2,7),
     *    A(7,7), S(7)
      REAL B(7), XBAR, YPOL, XMIN, XMAX, XP
      INTEGER N, KPLUS1, NROWS, IFAIL, J, K, I
      INTEGER LPLUS1, L, M, IP
```

continued

```
OPEN (UNIT=20,FILE='E02ADF.DAT')

NROWS = 7

WRITE (6,*) 'THE NUMBER OF POINTS'
READ (20,*) N
WRITE (6,*) N
WRITE (6,*) 'THE (X,Y) VALUES AND THEIR WEIGHTS'
DO 10 J = 1, N
   READ (20,*) X(J), Y(J), W(J)
   WRITE (6,*) X(J), Y(J), W(J)
10 CONTINUE
WRITE (6,*) 'THE HIGHEST DEGREE POLYNOMIAL REQUIRED'
READ (20,*) K
WRITE (6,*) K
KPLUS1 = K + 1

IFAIL = 0

CALL E02ADF(N,KPLUS1,NROWS,X,Y,W,WORK1,WORK2,A,S,
*              IFAIL)

DO 20 I = 1, K + 1
   IP = I - 1
   WRITE (6,*)
   WRITE (6,*)
*    'CHEBYSHEV COEFFS FOR POLYNOMIAL OF DEGREE ', IP
   WRITE (6,*) (A(I,J),J=1,I)
   WRITE (6,*) 'WITH RMS RESIDUAL ', S(I)
20 CONTINUE
WRITE (6,*)
WRITE (6,*)
* 'ENTER THE DEGREE OF YOUR BEST FIT POLYNOMIAL'
READ (5,*) L
LPLUS1 = L + 1
DO 30 I = 1, L + 1
   B(I) = A(L+1,I)
30 CONTINUE
XMIN = X(1)
XMAX = X(N)
WRITE (6,*) 'ENTER THE NUMBER OF Y VALUES REQUIRED'
READ (5,*) M
DO 40 I = 1, M
   WRITE (6,*) 'ENTER A VALUE FOR X BETWEEN ', XMIN,
*      ' AND ', XMAX
   READ (5,*) XP
   XBAR = ((XP-XMIN)-(XMAX-XP))/(XMAX-XMIN)

   CALL E02AEF(LPLUS1,B,XBAR,YPOL,IFAIL)

   WRITE (6,*) 'AT X = ', XP,
*      ', THE POLYNOMIAL GIVES Y = ', YPOL
40 CONTINUE

STOP

END
```

```
THE NUMBER OF POINTS
9
THE (X,Y) VALUES AND THEIR WEIGHTS
-4.200000, 43.00000, 1.000000
-2.800000, 24.00000, 1.000000
-2.000000, 9.200000, 1.000000
-1.300000, 2.100000, 1.000000
0.0000000E+00, 1.100000, 1.000000
0.8000000, 6.100000, 1.000000
2.200000, 16.00000, 1.000000
3.000000, 32.00000, 1.000000
4.100000, 58.00000, 1.000000
THE HIGHEST DEGREE POLYNOMIAL REQUIRED
4

CHEBYSHEV COEFFS FOR POLYNOMIAL OF DEGREE 0
42.55556
WITH RMS RESIDUAL 19.75404

CHEBYSHEV COEFFS FOR POLYNOMIAL OF DEGREE 1
42.45842, 7.256409
WITH RMS RESIDUAL 20.46190

CHEBYSHEV COEFFS FOR POLYNOMIAL OF DEGREE 2
52.14209, 5.627827, 24.89593
WITH RMS RESIDUAL 2.715458

CHEBYSHEV COEFFS FOR POLYNOMIAL OF DEGREE 3
52.05722, 5.754294, 24.97493, 1.310750
WITH RMS RESIDUAL 2.610052

CHEBYSHEV COEFFS FOR POLYNOMIAL OF DEGREE 4
52.20231, 5.718202, 25.06160, 1.303564, -0.4853010
WITH RMS RESIDUAL 2.859159

ENTER THE DEGREE OF YOUR BEST FIT POLYNOMIAL
 2
ENTER THE NUMBER OF Y VALUES REQUIRED
 4
ENTER A VALUE FOR X BETWEEN -4.200000 AND 4.100000
 -4.1
AT X = -4.100000, THE POLYNOMIAL GIVES Y = 43.10406
ENTER A VALUE FOR X BETWEEN -4.200000 AND 4.100000
 -3.0
AT X = -3.000000, THE POLYNOMIAL GIVES Y = 22.33435
ENTER A VALUE FOR X BETWEEN -4.200000 AND 4.100000
 -1.0
AT X = -1.000000, THE POLYNOMIAL GIVES Y = 2.496033
ENTER A VALUE FOR X BETWEEN -4.200000 AND 4.100000
 2.7
AT X = 2.700000, THE POLYNOMIAL GIVES Y = 26.76829
```

Note: Before you enter the degree of your polynomial of best fit you should have a look at the residuals which have been printed. In the case above, the residual drops from 20.5 for degree 1, to 2.1 for degree 2 and then does not substantially decrease. So it would be reasonable to choose a polynomial of degree 2 to fit your data.

6.7 Minimum of a function: E04JAF [A difficult routine]

The **purpose** of this routine is to find a local minimum of a function $f(x_1, x_2, \ldots, x_n)$, where the x-variables may be bounded

i.e. $l_j \leqslant x_j \leqslant u_j$ $(j = 1, 2, \ldots, n)$

where l_j and u_j are given numbers.

Comments:

(i) This routine attempts to find a *local* minimum of the function, but not necessarily a *global* minimum. The distinction is best illustrated by a graph of a function of one variable:

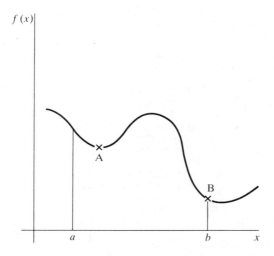

In the figure above, A is a local minimum in the interval $[a,b]$, but B is the global (or overall) minimum in this interval.

(ii) This routine can also be used to find a local maximum of a function f, by reformulating the problem, so that you minimize the function $-f$. For instance, if you wanted to maximize the function

$$f(x_1, x_2) = x_1^2 - \cos x_1 x_2 ,$$

then you would use the routine to minimize the function

$$g(x_1, x_2) = \cos x_1 x_2 - x_1^2 .$$

Note that the values of x_1 and x_2 which minimize g are the same as those which maximize f, and that the maximum value of f is minus the minimum value of g.

(iii) This routine may be used to minimize a function, where some, all or none of the variables $x_1, x_2, \ldots, x_n$ are bounded. However, the description given here requires you to provide bounds for all the variables, whether they are bounded or not. So, you are required to provide values for l_j and u_j for each variable so that $l_j \leqslant x_j \leqslant u_j$.

(a) If any bounds in the problem have the form

$$x_j \leqslant u_j ,$$

then these can always be rewritten

$$-\infty \leqslant x_j \leqslant u_j .$$

In this case, if you set $l_j = -1.0E6$, then the routine treats this value as $-\infty$.

(b) Similarly, if any bounds in the problem have the form

$$x_j \geqslant l_j ,$$

then these can be rewritten

$$l_j \leqslant x_j < \infty .$$

In this case, you should set $u_j = 1.0E6$.

(c) In the event that x_j is not bounded, then

$$-\infty \leqslant x_j \leqslant \infty .$$

In this case, you should set $l_j = -1.0E6$ and $u_j = 1.0E6$.

Specimen problem

To find a minimum of a function f given by

$$f(x_1, x_2, x_3, x_4) = (x_1 + 10x_2)^2 + 5(x_3 - x_4)^2 + (x_2 - 2x_3)^4 + 10(x_1 - x_4)^4 ,$$

subject to the constraints

$$1 \leqslant x_1 \leqslant 3$$
$$-2 \leqslant x_2$$
$$1 \leqslant x_4 \leqslant 3 .$$

Comments:

(i) Note that the variables do not all have to be bounded. In this example, x_3 is unbounded.

(ii) The routine requires you to make an initial guess at the solution (i.e. the x-values which minimize f). An initial guess at the x-values in the specimen problem could be $x_1 = 3.0$, $x_2 = -1.0$, $x_3 = 0.0$ and $x_4 = 1.0$. Note that these values are consistent with the given bounds on the variables.

The **method** uses a quasi-Newton algorithm.

The **routine name** with parameters is

E04JAF(N, IBOUND, BL, BU, X, F, IW, LIW, W, LW, IFAIL).

Description of parameters

Parameters which require values before E04JAF is called

N: [an integer variable]
N specifies the number of independent variables in the problem.

In the specimen problem, there are 4 independent variables x_1, x_2, x_3, and x_4, so N should be given the value 4.

Comment:
$N \geqslant 1$.

Ref. p. 19

X: [a real one-dimensional array. Its length must be at least N in the REAL declaration in your calling program]
The array X should contain an initial guess for the values of $x_1, x_2, \ldots, x_n$ which minimize the function f with the given constraints.

In the case of the specimen problem

$$X(1) = 3.0, \quad X(2) = -1.0, \quad X(3) = 0.0, \quad X(4) = 1.0 .$$

Comment:
The initial guess must be consistent with the bounds on the variables in the problem.

IBOUND: [an integer variable]
IBOUND must be given the value 0, 1, 2, or 3. If you set IBOUND = 0, then the routine expects all the x-variables to be bounded in the form

$$l_j \leqslant x_j \leqslant u_j .$$

The way in which this can be done is described in the introductory comment.

A fuller description of this parameter can be found in the *NAG Manual*.

BL ⎫
BU ⎭ : [real one-dimensional arrays. They must both have length at least N in the REAL declaration in your calling program]

The arrays BL and BU should contain the lower and upper bounds respectively on each variable $x_1, x_2, \ldots, x_n$.

In this case of the specimen problem, first the bounds should be written in the form

$$1 \leqslant x_1 \leqslant 3$$
$$-2 \leqslant x_2 \leqslant \infty$$
$$-\infty \leqslant x_3 \leqslant \infty$$
$$1 \leqslant x_4 \leqslant 3.$$

Hence, BL(1) = 1.0,   BU(1) = 3.0
       BL(2) = −2.0,   BU(2) = 1.0E6
       BL(3) = −1.0E6, BU(3) = 1.0E6
       BL(4) = 1.0,   BU(4) = 3.0.

Subroutines which require definition

The subroutine FUNCT1 described below is called by E04JAF.

Comment:

This subroutine is *not* a parameter of E04JAF, and so *must* be given the name FUNCT1. You must supply this routine at the end of your calling program.

FUNCT1: [a subroutine]

The **purpose** of this routine is to define the function f whose minimum is required.

The **name** of this routine with parameters is

        FUNCT1 (N, X, F) .

The way to use this subroutine and its parameters is best demonstrated by example.

If you wanted to write a subroutine which defined the function in the specimen problem, then a suitable subroutine FUNCT1 could be written as follows:

```
SUBROUTINE FUNCT1 (N, X, F)
INTEGER N
REAL X(N), F
F = (X(1) + 10.0*X(2))**2 + 5.0*(X(3) - X(4))**2 +
*    (X(2) - 2.0*X(3))**4 + 10.0*(X(1) - X(4))**4
RETURN
END
```

Comments on the subroutine FUNCT1

(i) The statement beginning 'F = . . .', in the subroutine above, effectively defines the function to be minimized.

(ii) Note that no values are assigned to N or to the array X in this subroutine. These values are supplied by the NAG routine E04JAF. *In no circumstances should you attempt to assign values to N or X in FUNCT1.*

(iii) The declaration

REAL X(N)

is allowable in this subroutine, as N is a parameter.

Parameters associated with workspace

IW: [an integer one-dimensional array. Its length must be at least (N + 2) in the INTEGER declaration in your calling program]
This array is used as workspace.

Ref. §3.3

W: [a real one-dimensional array. Its length must be at least 12N + $N(N - 1)/2$ (or 13 if N = 1) in the REAL declaration in your calling program]
This array is used as workspace.

LIW ⎱ [integer variables]
LW ⎰ LIW and LW must specify the lengths of IW and W respectively as declared in your calling program.

So, looking at the specimen problem, where N = 4,

$$12N + N(N - 1)/2 = 54 .$$

Thus, *minimum* declarations

REAL W(54)
and   INTEGER IW(6)

would be needed in your calling program. Also, the statements

LW = 54
and   LIW = 6

would also be needed.

The error parameter

IFAIL: [an integer variable]
IFAIL is the error parameter described in §3.2. It is recommended that you set

> IFAIL = 1

before you call E04JAF. Then after the routine is called, you must print or test the value of IFAIL in your program. If IFAIL = 0 after a call of the routine, then the routine was certainly able to do the specified problem. However, if the routine fails or if there is any uncertainty about the solution, then one of the following error or warning messages will be printed:

Error message	Meaning	Advice
IFAIL = 1	*Either* N < 1 *or* IBOUND ≠ 0, 1, 2, or 3 *or* some lower bound BL(J) has been specified greater than the corresponding upper bound BU(J) *or* LIW < N + 2 *or* LW < 13 if N = 1 *or* LW < 12N + N(N − 1)/2.	*Check* you have specified N, IBOUND, LIW and LW correctly – taking into account the restrictions on them. [Check your declaration of W and IW.] Check that each value of BL(J) is less than BU(J).
IFAIL = 2	Slow, or no convergence.	Try starting again using the current value in the array X as your initial guess. Get some help: the function specified may not have a minimum.
IFAIL = 3	The routine cannot guarantee that it has found a minimum.	Get some help. See comment (i) below.
IFAIL = 4	(Unlikely.)	The same as for IFAIL = 2.
IFAIL = 5 IFAIL = 6 IFAIL = 7 IFAIL = 8	There is increasing doubt as to whether the result found is in fact a minimum. [The doubt increases with the increasing value of IFAIL!]	Get some help. See comment (i).
IFAIL = 9	*Either* the problem has no finite solution, *or* FUNCT1 has not been correctly specified.	Check your subroutine FUNCT1. Get some help.

Comments:

(i) In the case of the two failure messages IFAIL = 3 and IFAIL = 5, the final value in $(x_1, x_2, . . ., x_n)$ may well give a good indication of the value of **x** which minimize the function. Unfortunately, if you set IFAIL = 0 before the routine is called, you will never be able to get these *x*-values printed. It is for this reason that you are advised to set IFAIL = 1 initially.

(ii) If in any doubt as to what to do when the routine fails, then get some expert help.

Parameters to be examined after calling E04JAF

F: [a real variable]
F will contain the lowest value of *f* that the routine has been able to find. If the routine has been successful, this will be a local minimum value of *f*.

X: [the one-dimensional array described previously]
Before the routine is called, X is used to hold an initial guess at the solution. However, after a successful call of the routine, the array X(1), X(2), . . ., X(N) will contain the required values of $x_1, x_2, . . ., x_n$ which give a minimum value of *f*.

Even if the routine has not been completely successful, this array will still hold the values of $x_1, x_2, . . ., x_n$ corresponding to the lowest value of *f* that the routine can find. This information can often be useful.

Specimen program

Program planning

1. *Declare*   REAL BL( ), BU( ), X( ), F, W( )
INTEGER N, IBOUND, IW( ), LIW, LW, IFAIL

2. *Set*   IBOUND, LIW, LW
   Read   N, BL, BU, X
   Set   IFAIL = 1

3. *Call*   E04JAF

4. *Print*   F, X

5. *Write routine* FUNCT1

Comments:

(i) The following program can be used to find a minimum value of a function *f* with up to 10 independent variables. Each time you want to minimize a new function you will have to write a new subroutine FUNCT1. If your function has more than 10 independent variables, you will have to alter the dimensions of the arrays in the declarations accordingly. You will also have to change the values given to LIW and LW in the program.

(ii) You may need to make some changes to this program in order to make it run correctly on your computer. *See §3.4 for details.*

E04JAF specimen program

```
C       E04JAF: MINIMUM OF A FUNCTION

        REAL BL(10), BU(10), X(10), F, W(165)
        INTEGER N, IBOUND, IW(12), LIW, LW, IFAIL, I

        IBOUND = 0
        LIW = 12
        LW = 165

        WRITE (6,*)
     *   'ENTER THE NUMBER OF INDEPENDENT VARIABLES'
        READ (5,*) N
        WRITE (6,*)
     *   'ENTER THE LOWER AND UPPER BOUNDS FOR EACH VARIABLE'
        WRITE (6,*) '(A NEW LINE FOR EACH VARIABLE)'
        DO 10 I = 1, N
           READ (5,*) BL(I), BU(I)
   10   CONTINUE
        WRITE (6,*) 'ENTER AN INITIAL GUESS FOR THE X-VALUES'
        READ (5,*) (X(I),I=1,N)

        IFAIL = 1

        CALL E04JAF(N,IBOUND,BL,BU,X,F,IW,LIW,W,LW,IFAIL)

        WRITE (6,*) 'IFAIL=', IFAIL
        WRITE (6,*) 'THE MINIMUM VALUE OF F IS '
        WRITE (6,*) F
        WRITE (6,*) 'WHICH OCCURS AT THE POINT '
        WRITE (6,*) '(', (X(I),I=1,N), ')'

        STOP

        END

        SUBROUTINE FUNCT1(N,X,F)
        INTEGER N
        REAL X(N), F
        F = (X(1)+10.0*X(2))**2 + 5.0*(X(3)-X(4))**2
     *      + (X(2)-2.0*X(3))**4 + 10.0*(X(1)-X(4))**4
        RETURN
        END
```

continued

E04JAF specimen run

```
ENTER THE NUMBER OF INDEPENDENT VARIABLES
4
ENTER THE LOWER AND UPPER BOUNDS FOR EACH VARIABLE
(A NEW LINE FOR EACH VARIABLE)
 1.0    3.0
-2.0    1.0E6
-1.0E6  1.0E6
 1.0    3.0
ENTER AN INITIAL GUESS FOR THE X VALUES
3.0  -1.0    0.0    1.0
IFAIL =0
THE MINIMUM VALUE OF F IS
2.433793
WHICH OCCURS AT THE POINT
(1.000000, -8.5462653E-02, 0.4092904, 1.000000)
```

Postscript

Comment:
In the specimen run, the routine returned IFAIL = 0, so you can be confident that it found a local minimum. There is no completely reliable method for checking whether this is also a global minimum; but if you think that the function might have other local minima within the bounds and want to find them, you should try further runs with different initial guesses at the solution X. If the function does have more than one local minimum, it will usually, though not always, converge to the one nearest to the initial guess.

Associated routines:
   (i) If the function which you want to minimize is linear in the variables $x_1, x_2, \ldots, x_n$, then use E04MBF, which is described in the next section. E04MBF also allows general linear constraints, i.e. it solves linear programming problems.
   (ii) If the function which you want to minimize can be expressed as a sum of squared terms, like the function f in the specimen problem, and there are no bounds on the variables, then use E04FDF which is described in §10.7.

Ref. §8   (iii) If you are likely to be solving several minimization problems, then you should consult the appropriate decision tree in the *NAG Mini-Manual*. This will help you to find the most suitable routine for your problem.

6.8 Linear programming: E04MBF
[A difficult routine]

The **purpose** of this routine is to find the x-values which minimize a linear function (called the *objective function*)

$$z = c_1x_1 + c_2x_2 + \ldots + c_nx_n$$

(where the c's are constant) subject to a set of linear constraints. These constraints may take the form either of *simple bounds* on the x-values, i.e.

$$l_1 \leqslant x_1 \leqslant u_1$$
$$l_2 \leqslant x_2 \leqslant u_2$$
.
.
.
$$l_n \leqslant x_n \leqslant u_n \, ,$$

or of *general linear constraints*, i.e.

$$l_{n+1} \leqslant a_{11}x_1 + a_{12}x_2 + \ldots + a_{1n}x_n \leqslant u_{n+1}$$
$$l_{n+2} \leqslant a_{21}x_1 + a_{22}x_2 + \ldots + a_{2n}x_n \leqslant u_{n+2}$$
.
.
.
$$l_{n+m} \leqslant a_{m1}x_1 + a_{m2}x_2 + \ldots + a_{mn}x_n \leqslant u_{n+m} \quad\quad (6.7)$$

Comments:

(i) This routine can also be used if you have a maximization problem. If you want to maximize a function then you can always reformulate the problem so that the function $-z$ is minimized.

For instance, if you wanted to maximize the function

$$z = 2x_1 - 3x_2 \, ,$$

then you would use the routine to minimize

$$y = -2x_1 + 3x_2 \, ,$$

Note that the values of x_1 and x_2 which minimize y are the same as those which maximize z, and that the maximum value of z is minus the minimum value of y.

(ii) The routine requires you to provide simple bounds on all the variables $x_1, x_2, \ldots, x_n$. So you are required to provide a value for l_j and u_j so that $l_j \leqslant x_j \leqslant u_j$ for each variable.

(a) If any of the bounds in the problem have the form

$$x_j \leqslant u_j \, ,$$

then these can always be rewritten

$$-\infty \leqslant x_j \leqslant u_j \, .$$

In this case, if you set $l_j = -1.0E20$, then the routine treats this value as $-\infty$.

(b) Similarly, if any bounds have the form

$$x_j \geqslant l_j \, ,$$

then these can be rewritten

$$l_j \leq x_j \leq \infty .$$

In this case, you should set $u_j = 1.0E20$.

(c) In the event that x is not bounded, then

$$-\infty \leq x_j \leq \infty .$$

In this case, you should set $l_j = -1.0E20$ and $u_j = 1.0E20$.

Similarly, if any of the general constraints of your problem have only a lower or upper bound, then you should also supply the 'missing' bound in the manner described above.

(iii) This routine can also be used just to find a point which satisfies the given constraints (i.e., a feasible point), rather than to solve the complete linear programming problem.

Specimen problem

To maximize the function

$$z = 5.2x_1 + 1.9x_2$$

with the simple bounds

$$\left. \begin{array}{r} x_1 \geq 0 \\ -1 \leq x_2 \leq 1 \end{array} \right\} ,$$

and 3 general linear constraints

$$\begin{array}{r} 2.1x_1 + 0.9x_2 \leq 9 \\ -1 \leq 1.0x_1 - 2.1x_2 \leq 1.9 \\ -3.1 \leq 2.9x_1 - 2.0x_2 \end{array}$$

The **method** is a specialized version of a general active-set method for linearly constrained minimization problems. In most circumstances it is equivalent to the well-known Simplex method. [See the *NAG Manual* for further details if needed.]

The **routine name** with parameters is

E04MBF(ITMAX, MSGLVL, N, M, NPLUSM, LA, A, BL, BU, C, LINOBJ, X, ISTATE, OBJLP, CLAMDA, IWORK, LIWORK, WORK, LWORK, IFAIL).

Description of parameters

Parameters which require values before E04MBF is called

N: [an integer variable]
N specifies the number of variables in the problem.

In the specimen problem above, there are two variables, x_1 and x_2, so N should be given the value 2.

Comments:
(i) $N \geqslant 1$.
(ii) N will also specify the number of simple bounds, as you have to specify simple bounds for each variable.

M: [an integer variable]
M specifies the number of general constraints in the problem.

In the specimen problem above, there are 3 general constraints, so M should be given the value 3.

Comment:
$M \geqslant 0$.

NPLUSM: [an integer variable]
You are asked to set this parameter to (M + N). A simple statement

NPLUSM = N + M

in your program deals with this parameter.

C: [a real one-dimensional array. Its length must be at least N in the REAL declaration in your calling program]
The array C is used to contain the coefficients $c_1, c_2, \ldots, c_n$ of the function which is to be minimized.

Ref.
p. 19

In the specimen problem, the problem is to maximize z. So, the problem is rewritten

Minimize $-z = -5.2x_1 - 1.9x_2$.

Hence, $C(1) = -5.2$ and $C(2) = -1.9$.

A: [a real two-dimensional array. It must have at least M rows and at least N columns in the REAL declaration in your calling program]
The array A is used to contain the coefficients of the general constraints (6.7). More specifically, the i(th) row of A contains the coefficients in the i(th) general constraint, and the j(th) column of A contains the coefficients of x_j.

In the case of the general constraints given in the specimen problem

$A(1, 1) = 2.1$ $\quad$ $A(1, 2) = 0.9$
$A(2, 1) = 1.0$ $\quad$ $A(2, 2) = -2.1$
$A(3, 1) = 2.9$ $\quad$ $A(3, 2) = -2.0$

Comment:
Even if no general linear constraints are being specified, an array A with at least one row and at least N columns must be supplied, although it will not be used.

LA: [an integer variable]

**Ref.
§2.4**
LA must specify the size of the first dimension of A as declared in your calling program.

For instance, if your specimen program had a declaration

REAL A(20, 30)

then a statement

LA = 20

would also be needed.

BL ⎫
BU ⎭ . [real one-dimensional arrays. Both must have length at least (M + N) in the REAL declaration in your calling program]

The arrays BL and BU are used to contain both the simple lower and upper bounds on the variables and the lower and upper bounds of the general linear constraints.

BL(1), BL(2), . . ., BL(N) hold the simple lower bounds on the N variables, and BL(N + 1), BL(N + 2), . . ., BL(N + M) hold the lower bounds for the M general linear constraints.

The array BU is set up in a similar way to contain both the simple upper bounds on the variables, and the upper bounds of the general linear constraints.

Comment:
Before this can be done, the constraints have to be arranged in the required form. For instance, the variables and the general linear constraints given in the specimen problem have to have both lower and upper bounds specified. Hence, these constraints could be rewritten

$$0 \leqslant x_1 \leqslant \infty$$
$$-1 \leqslant x_2 \leqslant 1$$
$$-\infty \leqslant 2.1x_1 + 0.9x_2 \leqslant 9.0$$
$$-1.0 \leqslant 1.0x_1 - 2.1x_2 \leqslant 1.9$$
$$-3.1 \leqslant 2.9x_1 - 2.0x_2 \leqslant \infty.$$

Now the constraints are in a form where the values can be assigned to BL and BU. In this case

$$BL(1) = \quad 0.0, \quad BU(1) = 1.0E20$$
$$BL(2) = -1.0, \quad BU(2) = 1.0$$
$$BL(3) = -1.0E20, \ BU(3) = 9.0$$
$$BL(4) = -1.0, \quad BU(4) = 1.9$$
$$BL(5) = -3.1, \quad BU(5) = 1.0E20.$$

X: [a real one-dimensional array. Its length must be at least N in the REAL declaration of your program]
Before the routine is called, $X(1), X(2), \ldots, X(N)$ must contain an initial guess at the solution. You are recommended initially to set them all to zero.

Comment:
It is not necessary for the initial guess to satisfy the constraints.

The remaining parameters described under this heading are of an administrative nature, but have to be assigned specific values before the routine is called.

ITMAX: [an integer variable]
ITMAX should specify the maximum number of iterations to be taken. Try starting with ITMAX = 50 and increase this only if necessary.

MSGLVL: [an integer variable]

If you set

MSGLVL = 0

then in general you will be left to print your own messages. In this event, the routine will print a message only if your problem is very ill-conditioned.

Other values of MSGLVL (message level) will give other levels of print-out, but read the *NAG Manual* before you use any other value.

LINOBJ: [a logical variable. This must be declared at the beginning of your calling program]

Ref.
§2.3

LINOBJ must be set to either .TRUE. or .FALSE. This parameter conveys to the routine whether you want to solve a linear programming problem or whether you just want to find a feasible point which satisfies the constraints.

Setting for LINOBJ	Meaning
LINOBJ = .TRUE.	A linear programming problem is to be solved.
LINOBJ = .FALSE.	A feasible point is required which satisfies the given constraints.

Comment
If you want just to find a point in the feasible region, then any values in the array C will be ignored. So C would not have to be specified in this instance, and the array X would contain an initial guess at a feasible point.

Parameters associated with workspace

IWORK: [an integer one-dimensional array. Its length must be at least 2N in the INTEGER declaration in your calling program]

Ref.
§3.3   This array is used as workspace.

WORK: [a real one-dimensional array. Its length must be at least K(*) in the REAL declaration in your calling program]
This array is used as workspace.

(*)Comment:
If the problem has general constraints specified, then

$$K = LA + 4M + 6N + 2I^2$$

where $I = \min(m+1, N)$.

However, if no general constraints have been specified, then

$$K = LA + 6N + 3.$$

So, in the case of the specimen problem where LA = 3, M (the number of constraints) = 3 and N (the number of variables) = 2, then

$$I = \min(4,2) = 2$$
and $$K = 3 + 12 + 12 + 2 \times 2^2 = 35.$$

Hence, a minimum declaration

REAL WORK(35)

would be needed in the calling program.

LIWORK⎫
LWORK ⎬ : [integer variables]

LIWORK and LWORK must specify the lengths of IWORK and WORK respectively as declared in your calling program. So, with the minimum declarations

REAL WORK(35)
INTEGER IWORK(4)

needed for the specimen problem, the statements

LWORK = 35
and   LIWORK = 4

would also be needed.

The error parameter

   IFAIL: [an integer variable]
   IFAIL is the error parameter described in §3.2. It is recommended that you set

   $$IFAIL = 0$$

   before you call E04MBF. Then in the event of the routine failing, your program will stop and print one of the following error messages:

Error message	Meaning	Advice
IFAIL = 1	No feasible point could be found.	Either the problem has no solution or there is a mistake in your data
IFAIL = 2	The linear programming problem has no finite solution.	
IFAIL = 3	The values of $x_1, x_2, \ldots,$ x_n are not changing.	Get advice as to how to use E04NAF with MSGLVL $\geq$ 5.
IFAIL = 4	The limit on the number of iterations has been reached.	Try increasing ITMAX. If that doesn't work, the advice is the same as for IFAIL = 3.
IFAIL = 5	Something is wrong with your parameters.	Check the values you gave to the input parameters, and their declarations and restrictions.

Comment:
   If in doubt as to what to do when the routine fails, get some expert help.

Parameters to be examined after calling E04MBF

   OBJLP: [a real variable]
   If all goes well in the routine, then OBJLP will contain the calculated minimum value of the objective function z.

   X: [the real one-dimensional array described previously]
   Before E04MBF is called, the array X is used to hold a guess at the solution. However, after the routine is called, then X(1), X(2), ..., X(N) contain the values $x_1, x_2, \ldots, x_n$ which give a minimum value of z. These values of x are the solution to the linear programming problem.

   ISTATE: [an integer one-dimensional array. Its length must be at least (N + M) in the declaration in your calling program]

This parameter gives information about each constraint at the final value of (X(1), X(2), . . ., X(N)). If you need information about the status of your constraints at the solution, then consult the *NAG Manual* for further details on this parameter.

CLAMDA: [a real one-dimensional array. Its length must be at least (N + M) in the REAL declaration in your calling program] CLAMDA contains the Lagrange multipliers (the final reduced costs), for both the simple bounds and the general linear constraints. Again, if you need more information, then consult the *NAG Manual* for further details.

Specimen program

Program planning

1. *Declare* REAL A(,), BL( ), BU( ), C( ), X( ), OBJLP, CLAMDA( ), WORK( )
INTEGER ITMAX, MSGLVL, N, M, NPLUSM, LA, ISTATE(,),IWORK( ), LIWORK, LWORK, IFAIL
LOGICAL LINOBJ

2. *Set*   MSGLVL, LA, LWORK, LIWORK, LINOBJ
Read N, C, M, A
Set   NPLUSM
Read BL, BU, X, ITMAX
Set   IFAIL

3. *Call*   E04MBF

4. *Print* OBJLP, X, (ISTATE, CLAMDA)

Comments:

(i) The following program can be used to find the minimum of a linear objective function which has up to 20 variables, subject to 20 general linear constraints. Linear programming problems can come in much larger sizes than this. If you want to use this program for a larger problem, then you will have to alter the dimensions of all the arrays in the declarations accordingly. You will also have to change the values given to LWORK, LIWORK, and LA in the program.

Ref.
§2.2
(ii) If you are entering large amounts of data, then you are recommended to use a data file.

(iii) You may need to make some changes to this program in order to make it run correctly on your computer. *See §3.4 for details.*

E04MBF specimen program

```
C     E04MBF: LINEAR PROGRAMMING

      REAL A(20,20), BL(40), BU(40), C(20), X(20)
      REAL OBJLP, CLAMDA(100), WORK(1020)
      INTEGER ITMAX, MSGLVL, N, M, NPLUSM, LA, ISTATE(40)
      INTEGER IWORK(40), LIWORK, LWORK, IFAIL, I, J
      LOGICAL LINOBJ

      MSGLVL = 0
      LA = 20
      LINOBJ = .TRUE.
      LIWORK = 40
      LWORK = 1020
      LINOBJ = .TRUE.

      WRITE (6,*) 'ENTER THE NUMBER OF VARIABLES'
      READ (5,*) N
      WRITE (6,*)
    * 'ENTER THE COEFFICIENTS OF THE OBJECTIVE FUNCTION'
      READ (5,*) (C(I),I=1,N)
      WRITE (6,*) 'ENTER THE NUMBER OF GENERAL CONSTRAINTS'
      READ (5,*) M
      WRITE (6,*)
    * 'ENTER THE COEFFICIENTS OF THE GENERAL CONSTRAINTS'
      WRITE (6,*) '(FOR EACH CONSTRAINT START A NEW LINE)'
      DO 10 I = 1, M
        READ (5,*) (A(I,J),J=1,N)
   10 CONTINUE
      NPLUSM = N + M
      WRITE (6,*) 'ENTER THE LOWER AND UPPER BOUNDS FOR'
      WRITE (6,*) 'EACH VARIABLE (ONE PAIR PER LINE)'
      DO 20 I = 1, N
        READ (5,*) BL(I), BU(I)
   20 CONTINUE
      WRITE (6,*) 'ENTER THE LOWER AND UPPER BOUNDS FOR'
      WRITE (6,*)
    * 'EACH GENERAL CONSTRAINT (ONE PAIR PER LINE)'
      DO 30 I = N + 1, N + M
        READ (5,*) BL(I), BU(I)
   30 CONTINUE
      WRITE (6,*) 'ENTER THE', N,
    * ' COORDINATES OF A GUESS AT THE SOLUTION'
      READ (5,*) (X(I),I=1,N)
      WRITE (6,*) 'ENTER THE MAXIMUM NUMBER OF ITERATIONS'
      READ (5,*) ITMAX

      IFAIL = 0
```

continued

```
      CALL E04MBF(ITMAX,MSGLVL,N,M,NPLUSM,LA,A,BL,BU,C,
    *            LINOBJ,X,ISTATE,OBJLP,CLAMDA,IWORK,LIWORK,
    *            WORK,LWORK,IFAIL)

   WRITE (6,*)
 * 'THE MINIMUM VALUE OF THE OBJECTIVE FUNCTION IS',
 * OBJLP
   WRITE (6,*) 'THIS MINIMUM OCCURS WHEN'
   DO 40 I = 1, N
      WRITE (6,*) 'X(', I, ') = ', X(I)
40 CONTINUE

   STOP

   END
```

E04MBF specimen run

```
ENTER THE NUMBER OF VARIABLES
 2
ENTER THE COEFFICIENTS OF THE OBJECTIVE FUNCTION
 -5.2   -1.9
ENTER THE NUMBER OF GENERAL CONSTRAINTS
 3
ENTER THE COEFFICIENTS OF THE GENERAL CONSTRAINTS
(FOR EACH CONSTRAINT START A NEW LINE)
 2.1    0.9
 1.0   -2.1
 2.9   -2.0
ENTER THE LOWER AND UPPER BOUNDS FOR
EACH VARIABLE (ONE PAIR PER LINE)
 0.0    1.0E20
 -1.0   1.0
ENTER THE LOWER AND UPPER BOUNDS FOR
EACH GENERAL CONSTRAINT (ONE PAIR PER LINE)
 -1.0E20 9.0
 -1.0    1.9
 -3.1    1.0E20
ENTER THE 2 COORDINATES OF A GUESS AT THE SOLUTION
 3.0    1.0
ENTER THE MAXIMUM NUMBER OF ITERATIONS
 50
THE MINIMUM VALUE OF THE OBJECTIVE FUNCTION IS  -21.97571
THIS MINIMUM OCCURS WHEN
X(1) =    3.881356
X(2) =    0.9435030
```

Comment:

Looking at the results above, you can see that the minimum of the specified objective function is -21.97571. So, going back to the original objective function, the maximum value is 21.97571, which occurs when $x_1 \simeq 3.881$ and $x_2 \simeq 0.944$.

6.9 Eigenvalues and eigenvectors: F02AGF
[A medium routine]

The **purpose** of this routine is to find the values of λ (the *eigenvalues*), and corresponding values of z (the *eigenvectors*) which satisfy the matrix equation

$$Az = \lambda z \quad (z \neq 0) .$$

Comment:
The square matrix A must be real, but the eigenvalues and eigenvectors may be complex.

Specimen problem

To find the eigenvalues and eigenvectors of the real 4×4 matrix

$$A = \begin{bmatrix} 4 & -5 & 0 & 3 \\ 0 & 4 & -3 & -5 \\ 5 & -3 & 4 & 0 \\ 3 & 0 & 5 & 4 \end{bmatrix}$$

The **method** reduces the matrix to upper Hessenberg form, and then uses the QR algorithm.

The **routine name** with parameters is

F02AGF (A, IA, N, RR, RI, VR, IVR, VI, IVI, INTGER, IFAIL).

Description of parameters

Parameters which require values before F02AGF is called

N: [an integer variable]
N is the size or order of the matrix A.

In the specimen problem, A is a 4×4 matrix, so N should be given the value 4.

A: [a real two-dimensional array. It must have at least N rows and at least N columns in the REAL declaration in your calling program]

**Ref.
p. 19**
The array A should contain the matrix whose eigenvalues and eigenvectors are required.

So, in the case of the specimen problem

$$A(1,1) = 4.0, \quad A(1,2) = -5.0, \quad A(1,3) = 0.0, \quad A(1,4) = 3.0$$
$$A(2,1) = 0.0, \quad A(2,2) = 4.0 \ldots, \text{ and so on}$$

IA: [an integer variable]

Ref.
§2.4 IA must specify the size of the first dimension of A as declared in your calling program. An example of how to set IA is given later.

Comment:
The following parameters fulfil a similar role to IA for the two arrays VR and VI. These arrays are described later.

$\left.\begin{array}{l} \text{IVR} \\ \text{IVI} \end{array}\right\}$: [integer variables]

IVR and IVI must specify the size of the first dimensions of the arrays VR and VI respectively as declared in your calling program.

So, if you had a (not very realistic) declaration

REAL A(10, 10), VR(8, 5), VI(7, 9)

in your calling program, then the statements

$$\begin{array}{rl} \text{IA} = & 10 \\ \text{IVR} = & 8 \\ \text{and} \quad \text{IVI} = & 7 \end{array}$$

would also be needed.

Parameters associated with workspace

INTGER: [an integer one-dimensional array. Its length must be at least N in

Ref.
§3.3 the INTEGER declaration in your calling program]
This array is primarily used as workspace.

The error parameter

IFAIL: [an integer variable]
IFAIL is the error parameter described in §3.2. It is recommended that you set

IFAIL = 0

before you call F02AGF. Then in the event of the routine failing, your program will stop and print the error message

IFAIL = 1.

This is the only error message for this routine. It means that the routine has failed to find all the eigenvalues. If this occurs, go and get some help.

Parameters to be examined after calling F02AGF

$\left.\begin{array}{l} \text{RR} \\ \text{RI} \end{array}\right\}$: [real one-dimensional arrays. Both these arrays must have length at least N in the REAL declaration in your calling program]
If all goes well in the routine, then the arrays RR and RI will contain the

real and imaginary parts respectively of the complex eigenvalues λ. So the eigenvalues λ_1, λ_2, . . ., λ_n will be found in $(RR(1), RI(1))$, $(RR(2), RI(2))$, . . ., $(RR(N), RI(N))$ respectively.

$\left.\begin{array}{l}VR \\ VI\end{array}\right\}$: [real two-dimensional arrays. Both these arrays must have at least N rows and at least N columns in the REAL declaration in your calling program] The arrays VR and VI contain the real and imaginary parts of the eigenvectors. More specifically, the real and imaginary parts of an eigenvector corresponding to the k(th) eigenvalue λ_k will be found in the k(th) *columns* of VR and VI respectively.

Thus, in the case of our specimen problem where A is a 4×4 matrix, the real and imaginary parts of the third eigenvector will be found in

$$
\begin{array}{ccc}
\begin{array}{c} VR(1, 3) \\ VR(2, 3) \\ VR(3, 3) \\ VR(4, 3) \end{array} & \text{and} & \begin{array}{c} VI(1, 3) \\ VI(2, 3) \\ VI(3, 3) \\ VI(4, 3) \end{array} & \text{respectively.}
\end{array}
$$

Specimen program

Program planning

1. *Declare* REAL A(,), RR( ), RI( ), VR(,), VI(,)
   INTEGER IA, N, IVR, IVI, INTGER( ), IFAIL

2. *Set*   IA, IVR, IVI
   Read N, A
   Set   IFAIL

3. *Call*   F02AGF

4. *Print* RR, RI, VR, VI

Comments:

(i) The following program can be used to find the eigenvalues and corresponding eigenvectors of a matrix of size up to 10×10. If you want to use the program for larger matrices, then you will have to alter the dimensions of the arrays in the declarations accordingly. You will also have to change the values given to IA, IVR, and IVI in the program.

(ii) You may need to make some changes to this program in order to make it run correctly on your computer. *See §3.4 for details.*

F02AGF specimen program

```
C       F02AGF: EIGENVALUES AND EIGENVECTORS
        REAL A(10,10), RR(10), RI(10), VR(10,10), VI(10,10)
        INTEGER IA, N, IVR, IVI, INTGER(10), IFAIL, J, K

        IA = 10
        IVR = 10
        IVI = 10

        WRITE (6,*) 'ENTER THE ORDER OF THE MATRIX'
        READ (5,*) N
        WRITE (6,*) 'ENTER THE MATRIX A, ONE ROW PER LINE'
        DO 10 J = 1, N
          READ (5,*) (A(J,K),K=1,N)
     10 CONTINUE

        IFAIL = 0

        CALL F02AGF(A,IA,N,RR,RI,VR,IVR,VI,IVI,INTGER,IFAIL)

        DO 30 K = 1, N
          WRITE (6,*)
          WRITE (6,*) 'AN EIGENVALUE IS'
          WRITE (6,*) RR(K), ' + ', RI(K), ' I'
          WRITE (6,*) 'WITH CORRESPONDING EIGENVECTOR'
          DO 20 J = 1, N
            WRITE (6,*) VR(J,K), ' + ', VI(J,K), ' I'
     20   CONTINUE
     30 CONTINUE

        STOP
```

F02AGF specimen run

```
ENTER THE ORDER OF THE MATRIX
4
ENTER THE MATRIX A, ONE ROW PER LINE
4.0   -5.0    0.0    3.0
0.0    4.0   -3.0   -5.0
5.0   -3.0    4.0    0.0
3.0    0.0    5.0    4.0

AN EIGENVALUE IS
12.00000 + 0.0000000E+00 I
WITH CORRESPONDING EIGENVECTOR
-0.5000000 + 0.0000000E+00 I
0.5000000 + 0.0000000E+00 I
-0.5000000 + 0.0000000E+00 I
-0.5000000 + 0.0000000E+00 I
```

```
AN EIGENVALUE IS
1.000000 + 5.000000 I
WITH CORRESPONDING EIGENVECTOR
0.0000000E+00 + 0.5000000 I
0.5000000 + 0.0000000E+00 I
0.5000000 + 5.8404575E-09 I
8.7606862E-09 + -0.5000000 I

AN EIGENVALUE IS
1.000000 + -5.000000 I
WITH CORRESPONDING EIGENVECTOR
0.0000000E+00 + -0.5000000 I
0.5000000 + 0.0000000E+00 I
0.5000000 + -5.8404575E-09 I
8.7606862E-09 + 0.5000000 I

AN EIGENVALUE IS
2.000000 + 0.0000000E+00 I
WITH CORRESPONDING EIGENVECTOR
0.5000000 + 0.0000000E+00 I
0.5000000 + 0.0000000E+00 I
-0.5000000 + 0.0000000E+00 I
0.5000000 + 0.0000000E+00 I
```

Postscript

Comments:

(i) Eigenvectors are not unique. If you run the specimen problem, you might well find that the eigenvectors which you obtain differ by some multiple from the ones in the specimen run. For example, using the same matrix, the eigenvectors obtained above are $1, -i, i$, and -1 times those in the *NAG Manual*. It is important to remember this when checking your results.

Ref. §5.2 (ii) If your matrix A is large, then you are recommended to use a data file.

Accuracy:

The results above are guaranteed to be reasonably accurate. Note that numbers like $5.8404575E-09$ are effectively zero on a machine with 8 digits of precision, and should be treated as such.

Associated routines:

(i) If your matrix is symmetric, then use F02ABF. This can be found in §10.2.

Ref. §8 (ii) If you are often likely to be using eigenvalue routines, or your matrix has special properties (such as the ones above), then you should consult the appropriate decision tree in the *NAG Mini-Manual*. This will help you choose the routine in the *NAG Manual* best suited to your problem.

6.10 Linear simultaneous equations: F04AEF
[A medium routine]

The **purpose** of this routine is to solve sets of n linear simultaneous equations in n unknowns.

Comments:

(i) In matrix terms, the routine finds the solution X of the matrix equation

$$AX = B.$$ (6.8)

where A and B are given matrices, with A square.

(ii) This routine is very flexible, and can be used in a number of different ways.

(a) The routine can be used to solve equations like

$$\left. \begin{array}{rcl} 2x_1 + x_2 &=& 3 \\ 3.7x_1 - 4.2x_2 &=& 8 \end{array} \right\}.$$

Equations such as these can always be expressed in matrix form,

i.e $\begin{bmatrix} 2 & 1 \\ 3.7 & -4.2 \end{bmatrix} \begin{bmatrix} x_1 \\ x_2 \end{bmatrix} = \begin{bmatrix} 3 \\ 8 \end{bmatrix}.$

So, in this instance, you would choose X and B to be 2×1 matrices with

$$X = \begin{bmatrix} x_1 \\ x_2 \end{bmatrix}, B = \begin{bmatrix} 3 \\ 8 \end{bmatrix} \text{ and A to be the } 2 \times 2 \text{ matrix } \begin{bmatrix} 2 & 1 \\ 3.7 & -4.2 \end{bmatrix}.$$

In general, if you want to solve n linear simultaneous equations (with just one right-hand side), then you would choose X to be an $n \times 1$ matrix and B to be the $n \times 1$ matrix containing the right-hand side.

Often, an $n \times 1$ matrix is referred to as a column vector. So, for the use of the routine described above, finding the solution X in AX = B is equivalent to finding the solution $\mathbf{x}$ in $A\mathbf{x} = \mathbf{b}$, where $\mathbf{x}$ and $\mathbf{b}$ are column vectors.

(b) Suppose you wanted to solve three sets of equations

$$\begin{bmatrix} 2 & 1 \\ 3.7 & -4.2 \end{bmatrix} \mathbf{x}_k = \mathbf{b}_k$$

where $\mathbf{b}_1 = \begin{bmatrix} 3 \\ 8 \end{bmatrix}$, $\mathbf{b}_2 = \begin{bmatrix} 5 \\ 7 \end{bmatrix}$, $\mathbf{b}_3 = \begin{bmatrix} 0 \\ 1 \end{bmatrix}$ and $\mathbf{x}_1$, $\mathbf{x}_2$, and $\mathbf{x}_3$ are three unknown column vectors.

These problems can be put together as one matrix equation

$$\begin{bmatrix} 2 & 1 \\ 3.7 & -4.2 \end{bmatrix} [\mathbf{x}_1, \mathbf{x}_2, \mathbf{x}_3] = \begin{bmatrix} 3 & 5 & 0 \\ 8 & 7 & 1 \end{bmatrix}$$
$$= [\mathbf{b}_1, \mathbf{b}_2, \mathbf{b}_3].$$

Thus, these three problems can be treated as a single matrix problem – namely, to solve.

$$AX = B,$$

where A is a 2×2 matrix, and X (the unknown matrix) and B are 2×3 matrices.

In general, if you want to solve m sets of n linear simultaneous equations, with the same left-hand side coefficients in each case but different right-hand sides, then you choose the matrices X and B in (6.8) to be $n \times m$ matrices.

(c) If you wish to find the inverse of an $n \times n$ matrix A, then you can choose the matrix B to be the $n \times n$ unit matrix I, and find the solution to the matrix equation

$$AX = I.$$

So, as $X = A^{-1}$, the resulting solutions $x_1, x_2, \ldots, x_n$ together give the inverse matrix A^{-1}.

Specimen problem

To solve the equations

$$\begin{aligned}
33x_1 + 16x_2 + 72x_3 &= -359 \\
-24x_1 - 10x_2 - 57x_3 &= 281 \\
- 8x_1 - 4x_2 - 17x_3 &= 85
\end{aligned}$$

Writing these equations in matrix form $AX = B$ then

$$A = \begin{bmatrix} 33 & 16 & 72 \\ -24 & -10 & -57 \\ -8 & -4 & -17 \end{bmatrix}, X = \begin{bmatrix} x_1 \\ x_2 \\ x_3 \end{bmatrix} \text{ and } B = \begin{bmatrix} -359 \\ 281 \\ 85 \end{bmatrix}.$$

The **method** decomposes A using Crout's factorization, followed by forward and backward substitution and iterative refinement.

The **routine name** with parameters is

F04AEF(A,IA,B,IB,N,M,X,IX,W,AA,IAA,BB,IBB,IFAIL).

Description of parameters

Parameters which require values before F04AEF is called

N: [an integer variable]

N is the number of equations to be solved, or the number of rows in the matrix A in (6.8).

In the specimen problem, there are 3 equations, so N should be given the value 3.

**Ref.
p. 19**

A: [a real two-dimensional array. It must have at least N rows and at least N columns in the REAL declaration in your calling program]
The array A is used to contain the coefficients of the left-hand sides of your equations.

Thus, in the case of the specimen problem

$$A(1,1) = \quad 33.0, \quad A(1,2) = 16.0, \quad A(1,3) = 72.0$$
$$A(2,1) = -24.0, \ldots \text{ and so on.}$$

**Ref.
§2.4**

IA: [an integer variable]
IA must specify the size of the first dimension of A as declared in your calling program. An example of how to set IA is given later.

M: [an integer variable]
M is the number of different right-hand sides.

In the specimen problem, there is just one right-hand side, so M should be given the value 1.

B: [a real two-dimensional array. It must have at least N rows and at least M columns in the REAL declaration in your calling program]
Each column of B is used to contain a right-hand side of the set of equations. So the k(th) column of B will hold the k(th) right-hand side.

In the specimen problem, there is only one right-hand side, so

$$B(1,1) = -359.0$$
$$B(2,1) = \quad 281.0$$
$$\text{and} \quad B(3,1) = \quad \quad 85.0.$$

IB: [an integer variable]
IB must specify the size of the first dimension of B as declared in your calling program.

Comment:
The following parameters fulfil a similar role to IA and IB for the three arrays X, AA and BB. These arrays are described later.

IX
IAA : [integer variables]
IBB

IX, IAA and IBB must specify the size of the first dimensions of the arrays X, AA and BB respectively as declared in your calling program.

So, if you had a (not very realistic) declaration

$$\text{REAL } A(5,4), B(6,3), X(7,5), AA(5,10), BB(8,12)$$

in your calling program, then the statements

$$IA = 5$$
$$IB = 6$$
$$IX = 7$$
$$IAA = 5$$
and $$IBB = 8$$

would also be needed.

Parameters associated with workspace

W: [a real one-dimensional array. Its length must be at least N in the REAL

Ref.   declaration in your calling program]
§3.3   This array is used as workspace.

The error parameter

IFAIL: [an integer variable]
IFAIL is the error parameter described in §3.2. It is recommended that you set

$$IFAIL = 0$$

before you call F04AEF. Then in the event of the routine failing, your program will stop and print one of the following error messages:

Error message	Meaning	Advice
IFAIL = 1	The matrix is singular, possibly due to rounding errors.	If your matrices A and B come from a physical problem then you may have supplied repeated information. Get some expert help.
IFAIL = 2	The problem is ill-conditioned.	

Parameters to be examined after calling F04AEF

X: [a real two-dimensional array. It must have at least N rows and at least M columns in the REAL declaration in your calling program]
If all goes well in the routine, then the solution of the k(th) set of equations (i.e. the solution using the k(th) column of B) will be found in the k(th) column of the array X.

Thus, in the specimen problem, where there is just one set of equations to be solved, the solution will be found in X(1,1), X(2,1), and X(3,1) (i.e. the first column of X).

Comment:
The following parameters may be of interest if you are checking your matrix for ill-conditioning. If you need some help, then get the following

arrays printed first. *In any case, they must be declared in your calling program.*

AA: [a real two-dimensional array. It must have at least N rows and at least N columns in the REAL declaration in your calling program] The array AA will contain the Crout factorization LU of A. The 1's on the diagonal of U are suppressed.

BB: [a real two-dimensional array. It must have at least N rows and at least M columns in the REAL declaration in your calling program] This array contains the M residual vectors corresponding to the different solutions.

Comment:
You will find further information about the Crout factorization and residual vectors in the *NAG Manual*, if required.

Specimen program

Program planning

1. *Declare* REAL A(,), B(,), X(,), W( ), AA(,), BB(,)
   INTEGER IA, IB, N, M, IX, IAA, IBB, IFAIL

2. *Set*   IA, IB, IX, IAA, IBB
   Read   N, M, A, B
   Set   IFAIL

3. *Call*   F04AEF

4. *Print* X (and AA, BB if necessary)

Comments:
(i) The following program can be used to solve sets of up to 20 equations in 20 unknowns. Up to 5 right-hand sides can be entered. If you want to use the program for more equations or more right-hand sides, then you will have to alter the dimensions of the arrays in the REAL declaration accordingly. You will also have to change the values given to IA, IB, IX, IAA, and IBB in the program.

(ii) You may need to make some changes to this program in order to make it run correctly on your computer. *See §3.4 for details.*

F04AEF specimen program

```
C       F04AEF: LINEAR SIMULTANEOUS EQUATIONS
C               (WITH MORE THAN ONE RIGHT-HAND SIDE)
```

```
      REAL A(20,20), B(20,5), X(20,20), W(20), AA(20,20),
     *      BB(20,5)
      INTEGER IA, IB, N, M, IX, IAA, IBB, IFAIL, I, J

      IA = 20
      IB = 20
      IX = 20
      IAA = 20
      IBB = 20

      WRITE (6,*) 'ENTER THE NUMBER OF EQUATIONS'
      READ (5,*) N
      WRITE (6,*) 'ENTER THE NUMBER OF RIGHT HAND SIDES'
      READ (5,*) M
      WRITE (6,*) 'ENTER THE MATRIX A'
      DO 10 I = 1, N
        READ (5,*) (A(I,J),J=1,N)
   10 CONTINUE
      WRITE (6,*) 'ENTER THE MATRIX B'
      WRITE (6,*) '(EACH RIGHT-HAND SIDE IN A COLUMN)'
      DO 20 I = 1, N
        READ (5,*) (B(I,J),J=1,M)
   20 CONTINUE

      IFAIL = 0

      CALL F04AEF(A,IA,B,IB,N,M,X,IX,W,AA,IAA,BB,IBB,IFAIL)

      WRITE (6,*) 'THE SOLUTIONS (ONE IN EACH COLUMN) ARE'
      DO 30 I = 1, N
        WRITE (6,'(1X,5E15.4)') (X(I,J),J=1,M)
   30 CONTINUE

      STOP

      END
```

F04AEF specimen run

```
ENTER THE NUMBER OF EQUATIONS
 3
ENTER THE NUMBER OF RIGHT HAND SIDES
 1
ENTER THE MATRIX A
  33.0    16.0    72.0
 -24.0   -10.0   -57.0
  -8.0    -4.0   -17.0
ENTER THE MATRIX B
(EACH RIGHT-HAND SIDE IN A COLUMN)
 -359.0
  281.0
   85.0
THE SOLUTIONS (ONE IN EACH COLUMN) ARE
     0.1000E+01
    -0.2000E+01
    -0.5000E+01
```

continued

Postscript

Comment:

(i) The output was 'formatted' so that the solutions would appear strictly in columns.

Ref. (ii) If your matrix A is large, then you are recommended to use a data
§5.2 file.

Accuracy:

The solutions should be correct to full machine accuracy.

Associated routines:

(i) If you have a large symmetric positive-definite matrix A, then use F04ASF. You will find the documentation for this in Chapter 9.

(ii) If you have a large symmetric positive-definite matrix A which has band structure, then use F04ACF which you will find in §10.6.

(iii) If your equations contain complex numbers, then use F04ADF which you will find in §10.3.

(iv) If you are often likely to be using routines to solve simultaneous equations, or if your matrix has special properties (such as the ones described above), then you should consult the appropriate decision tree in the *NAG Mini-Manual*. This will help you choose the routine in the *NAG Manual* best suited to your problem. This particular decision tree can be found in Chapter 8.

The following run demonstrates how this routine can be used to find the inverse of a given matrix. The matrix A in the specimen problem has been used for demonstration purposes. However, note that it is seldom necessary to find an inverse numerically.

```
ENTER THE NUMBER OF EQUATIONS
 3
ENTER THE NUMBER OF RIGHT HAND SIDES
 3
ENTER THE MATRIX A
 33.0    16.0    72.0
-24.0   -10.0   -57.0
 -8.0    -4.0   -17.0
ENTER THE MATRIX B
(EACH RIGHT-HAND SIDE IN A COLUMN)
 1.0   0.0   0.0
 0.0   1.0   0.0
 0.0   0.0   1.0
THE SOLUTIONS (ONE IN EACH COLUMN) ARE
-0.9667E+01     -0.2667E+01     -0.3200E+02
0.8000E+01      0.2500E+01      0.2550E+02
0.2667E+01      0.6667E+00      0.9000E+01
```

Comment:

The three solutions above, together give columns of the inverse matrix A.

6.11 Summary statistics: G01AAF
[An easy routine]

The **purpose** of this routine is to calculate the following statistics:

> the mean,
> the standard deviation,
> the coefficient of skewness,
> the coefficient of kurtosis,

and   the maximum and minimum of a given set of n observations.

If required, each observation may be weighted.

Comment:
You are reminded that these statistics do not automatically imply anything about the population from which the sample is drawn. Interpretation will depend on such matters as the degree to which the sample represents the population, and the population probability distribution.

Specimen problem

To calculate the statistics above for the 7 observations

> 14.2   15.3   16.4   12.1   18.2   21.1   17.4

with no weighting.

Method: The statistics above are calculated directly from standard formulae.

Comment:
The standard deviation calculated is the population estimate, which is obtained by dividing by $n - 1$, where n is the number of observations in the sample.

The **routine name** with parameters is

> G01AAF(N,X,IWT,WT,XMEAN,S2,S3,S4,XMIN,XMAX,WTSUM,
> IFAIL).

Description of parameters

Parameters which require values before G01AAF is called

N: [an integer variable]
N is the number of observations.

In the specimen problem, there are 7 observations, so N should be given the value 7 in this instance.

Ref.
p. 19

X: [a real one-dimensional array. Its length must be at least N in the REAL declaration in your calling program]
The array X is used to contain the observations.

So, in the case of the specimen problem, $X(1) = 14.2$, $X(2) = 15.3$, and so on.

IWT: [an integer variable]
IWT must either be set to 0 or 1 before the routine is called.

Setting for IWT	Meaning
IWT = 0	No weights supplied.
IWT = 1	User will supply weights.

Comment:
When all the observations are of equal importance, then they are referred to as 'not weighted', and IWT would be given the value 0. The specimen problem is an example of this.

WT: [a real one-dimensional array. Its length must be at least N in the REAL declaration in your calling program]
This array is used to contain the weights. WT(K) should contain the weight corresponding to the observation X(K).

Comments:
(i) If you want an observation X(K) in your data to be ignored, then the corresponding weight WT(K) should be given the value zero, and IWT the value 1.
(ii) Only the observations with non-zero weighting factors are used in the calculations. These observations are referred to as *valid observations*.
(iii) If IWT has been given the value zero, then there is no need to set WT(1), WT(2), . . ., WT(N) to 1. This is done automatically by the routine.

The error parameter
IFAIL: [an integer variable]
IFAIL is the error parameter described in §3.2. It is recommended that you set

IFAIL = 0

before you call G01AAF. Then in the event of the routine failing, your program will stop and print one of the following error messages:

Error message	Meaning	Advice
IFAIL = 1	N < 1.	Check your value of N.
IFAIL = 2	Only one valid observation was made (i.e. only one with a non-zero weighting).	Check the values in WT(1), WT(2), . . ., WT(N). You need more than 1 valid observation to calculate S2, S3, S4.
IFAIL = 3	*Either* there are no valid observations *or* a negative weight was specified.	Check the values in WT(1), WT(2), . . ., WT(N).

Parameters to be examined after calling G01AAF

XMIN
XMAX }: [real variables]

XMIN and XMAX will contain the minimum and maximum values respectively of the valid observations.

XMEAN: [a real variable]

XMEAN will contain the weighted mean value of the valid observations.

S2: [a real variable]

S2 will contain the standard deviation of the valid observations.

S3: [a real variable]

S3 will contain the coefficient of skewness of the valid observations.

S4: [a real variable]

S4 will contain the coefficient of kurtosis of the valid observations.

Comments:

(i) S2, S3 and S4 can be calculated only if there is more than one valid observation.

(ii) If you need to check your answer, you might find the following parameters useful at this stage.

IWT: [an integer variable]

Before G01AAF is called, IWT indicates whether weights were being supplied. However, after the routine is called, IWT will contain the number of valid observations.

WTSUM: [a real variable]

WTSUM will contain the sum of the weights. If your problem was unweighted initially, then WTSUM will contain the value N.

Specimen program

Program planning

1. *Declare* REAL X( ), WT( ), XMEAN, S2, S3, S4, XMIN, XMAX, WTSUM

   INTEGER N, IWT, IFAIL

2. *Read* N, X, IWT, WT (if necessary)

   Set IFAIL

3. *Call* G01AAF

4. *Print* XMIN, XMAX, XMEAN, S2, S3, S4

Comment:

(i) The following program can be used to find the specified statistics for up to 100 observations. If you have more observations than this, then you will have to alter the dimensions in the REAL declaration accordingly.

(ii) You may need to make some changes in this program in order to make it run correctly on your computer. *See §3.4 for details.*

G01AAF specimen program

```
C       G01AAF: DESCRIPTIVE STATISTICS
        REAL X(100), WT(100), XMEAN, S2, S3, S4, XMIN, XMAX,
     *       WTSUM
        INTEGER N, IWT, IFAIL, I

        WRITE (6,*) 'ENTER THE NUMBER OF OBSERVATIONS'
        READ (5,*) N
        WRITE (6,*) 'ENTER THE OBSERVATIONS, ONE PER LINE'
        DO 10 I = 1, N
          READ (5,*) X(I)
     10 CONTINUE
        WRITE (6,*)
     *  'ENTER 1 IF THERE ARE WEIGHTS, 0 OTHERWISE'
        READ (5,*) IWT
        IF (IWT.EQ.1) THEN
          WRITE (6,*) 'ENTER THE WEIGHTS, ONE PER LINE'
          DO 20 I = 1, N
            READ (5,*) WT(I)
     20   CONTINUE
        END IF

        IFAIL = 0

        CALL G01AAF(N,X,IWT,WT,XMEAN,S2,S3,S4,XMIN,XMAX,WTSUM,
     *              IFAIL)
```

```
      WRITE (6,*) 'THE MINIMUM AND MAXIMUM VALUES'
      WRITE (6,*) XMIN, XMAX
      WRITE (6,*) 'THE MEAN VALUE'
      WRITE (6,*) XMEAN
      WRITE (6,*) 'THE STANDARD DEVIATION'
      WRITE (6,*) S2
      WRITE (6,*)
    *  'THE COEFFICIENTS OF SKEWNESS AND KURTOSIS'
      WRITE (6,*) S3, S4

      STOP

      END
```

G01AAF specimen run

```
ENTER THE NUMBER OF OBSERVATIONS
 7
ENTER THE OBSERVATIONS, ONE PER LINE
 14.2
 15.3
 16.4
 12.1
 18.2
 21.1
 17.4
ENTER 1 IF THERE ARE WEIGHTS, 0 OTHERWISE
 0
THE MINIMUM AND MAXIMUM VALUES
12.10000, 21.10000
THE MEAN VALUE
16.38571
THE STANDARD DEVIATION
2.911717
THE COEFFICIENTS OF SKEWNESS AND KURTOSIS
0.1441387, -0.9887403
```

Postscript

Comment:

This routine could be used to calculate statistics for data which are given in the form of a frequency distribution. In this case, the array X would contain the N different values that the observations could take, and the array WT would contain the corresponding frequencies. In this event, WTSUM (*not* IWT) will contain the total number of observations after the routine is called.

Associated routines:

If you want to find the median, the quartiles (or other quantiles) for your observations, then you could make use of a sort routine such as M01AKF. This is described in §6.14.

6.12 Contingency table: G01AFF
[A medium/difficult routine]

The **purpose** of this routine is to test the assumption of independence of the variables in a two-way contingency table.

Specimen problem

(a) Test the assumption that credit-rating is independent of size of firm, given the following two-way table of frequencies:

		Size of firm		
		Large	Medium	Small
Credit	Good	28	60	57
rating	Poor	12	44	53

(b) Test the assumption that cure is independent of use of drug, given the table:

	Drug	No drug
Cure	11	2
No cure	3	9

Method: Normally, a chi-squared statistic is calculated. However, in the case of a 2×2 table with fewer than 41 observations (such as in the second table above), the probabilities for Fisher's exact test are calculated instead. For details see S. Siegel, *Nonparametric Statistics for the Behavioural Sciences*, pp. 95–111.

The **routine name** with parameters is

G01AFF(INOB,IPRED,M,N,NOBS,NUM,PRED,CHIS,P,NPOS,
NDF,M1,N1,IFAIL).

Description of parameters

Parameters which require values before G01AFF is called

M
N : [integer variables]

Suppose there are NR rows and NC columns in the contingency table. Then

M = NR + 1
and   N = NC + 1.

For example, the contingency table in specimen problem (a) has 2 rows and 3 columns. So, M and N should be given the values 3 and 4 respectively.

NOBS: [an integer two-dimensional array. It must have at least M rows and at least N columns in the INTEGER declaration in your calling program] NOBS (I,J) is used to contain the frequency in the i(th) row and j(th) column of the contingency table.

Ref.
p. 19

So, for the contingency table given in specimen problem (a),

$$NOBS(1,1) = 28, \quad NOBS(1,2) = 60, \quad NOBS(1,3) = 57,$$
$$NOBS(2,1) = 12, \quad NOBS(2,2) = 44, \quad NOBS(2,3) = 53.$$

Comment:
The extra row and column in this array is used after the routine is called.

INOB
IPRED $\Big\}$: [integer variables]

Ref.
§2.4

INOB and IPRED must specify the size of the first dimensions of the arrays NOBS and PRED respectively as they appear in the declaration in your calling program. (The array PRED will be described later.)

So, if there were the (not very realistic) declarations

REAL PRED (6, 8)
INTEGER NOBS (7, 8)

in your calling program, then the statements

IPRED = 6
and   INOB   = 7

would also be needed.

NUM: [an integer variable]
NUM must be set to either 0 or 1 before the routine is called.

Setting for NUM	Meaning
NUM = 0	No shrinkage required.
NUM = 1	Automatic shrinkage required.

In this description it is assumed for simplicity that no shrinkage is required, so NUM should be set to 0. For details of automatic shrinkage, see the *NAG Manual*.

The error parameter

IFAIL: [an integer variable]

IFAIL is the error parameter described in §3.2. It is recommended that you set

IFAIL = 0

before you call G01AFF. Then in the event of the routine failing, your program will stop and print one of the following error messages:

Error message	Meaning	Advice
IFAIL = 1	NOBS has less than 2 rows or columns.	Check whether you have entered the problem correctly.
IFAIL = 2	*Either* all the frequencies are zero *or* at least one frequency is negative.	
IFAIL = 3	Error inside routine.	Get some help.
IFAIL = 4	*Either* IPRED < M *or* INOB < M.	Check declaration of PRED, NOBS and values of IPRED, INOB, and M.

Parameters to be examined after calling G01AFF

M1
N1 } : [integer variables]

These parameters can be ignored except when shrinkage is requested.

NUM: [an integer variable]

After the routine is called, NUM has one of the following values:

Value assigned to NUM	Meaning
0	The chi-squared statistic has been calculated.
A positive integer K	Fisher's exact test is to be applied, using the probabilities $P(1), P(2), \ldots, P(K)$. (P is described below.)

So NUM can be tested to check which of the following parameters are to be printed for your particular problem.

The following parameters will be of interest if NUM = 0 after a routine call (i.e. if a chi-squared statistic is being calculated):

PRED: [a real two-dimensional array. It must have at least M rows and at least N columns in the REAL declaration in your calling program]

The array PRED will contain the expected values of the frequencies. In particular, PRED(I,J) will contain the expected frequency corresponding to the observed frequency NOBS(I,J).

CHIS: [a real variable]
CHIS will contain the value of the chi-squared statistic.

NDF: [an integer variable]
NDF will contain the number of degrees of freedom for use in the chi-squared test.

The following parameters will be of interest if NUM has a non-zero value after a routine call (i.e. if Fisher's exact test is to be used):

P: [a real one-dimensional array. Its length must be at least 21 in the declaration in your calling program]
The array P will contain the probabilities associated with all possible frequency tables with the given row and column totals (assuming that the variables are independent).

Comment:
The specimen table

11	2
3	9

is just one of the 12 possible tables of the form:

			Row totals
	$2 + x$	$11 - x$	13
	$12 - x$	x	12
Column totals	14	11	

with row totals 13 and 12, and column totals 14 and 11.

Note that x can take the 12 values $0, 1, \ldots, 11$. Also, that once x has been given a value, the rest of the table is fixed.

The probabilities of each of these tables can be found in the array P. Thus, P(1) contains the probability that $x = 0$, P(2) contains the probability that $x = 1$, and so on. The particular probability associated with the specimen table (i.e. when $x = 9$) is P(10).

However, as all the probabilities P(1), P(2), $\ldots$, P(12) are required in Fisher's exact test, they should all be printed.

₊NPOS: [an integer variable]

NPOS will contain a positive number L (say). This indicates that the element P(L) contains the probability associated with the observed frequency table.

Specimen program

Program planning

1. *Declare* REAL PRED(,), CHIS, P( )
   INTEGER INOB, IPRED, M, N, NOBS(,), NUM,
   NPOS, NDF, M1, N1, IFAIL

2. *Set*   IPRED, INOB, NUM
   Read   NR, NC (the number of rows and columns in the two-way contingency table)
   Set   M, N
   Read   NOBS
   Set   IFAIL

3. *Call*   G01AFF

4. *If* NUM > 0, *print* P, NPOS
   Else print PRED, CHIS, NDF

Comments:

(i) The following program can be used for up to a 10 × 10 two-way contingency table. If your table is larger than this, then you will have to alter the dimensions of PRED and NOBS in the declarations in the calling program accordingly. You will also have to change the values given to IPRED and INOB in the program.

(ii) You may need to make some changes to this program in order to make it run correctly on your computer. *See §3.4 for details.*

G01AFF specimen program

```
C       G01AFF: CONTINGENCY TABLE

        REAL PRED(11,11), CHIS, P(21)
        INTEGER INOB, IPRED, M, N, NOBS(11,11), NUM, NPOS,
     *          NDF, M1, N1, IFAIL, NR, NC, I, J

        INOB = 11
        IPRED = 11
        NUM = 0
```

```
      WRITE (6,*) 'ENTER THE NUMBER OF ROWS IN THE TABLE'
      READ (5,*) NR
      WRITE (6,*) 'ENTER THE NUMBER OF COLUMNS IN THE TABLE'
      READ (5,*) NC
      M = NR + 1
      N = NC + 1
      WRITE (6,*) 'ENTER TABLE FREQUENCIES ROW BY ROW'
      DO 10 I = 1, NR
        READ (5,*) (NOBS(I,J),J=1,NC)
   10 CONTINUE

      IFAIL = 0

      CALL G01AFF(INOB,IPRED,M,N,NOBS,NUM,PRED,CHIS,P,NPOS,
     *            NDF,M1,N1,IFAIL)

      IF (NUM.GT.0) THEN
        WRITE (6,*)
     *    'PROBABILITIES NEEDED FOR FISHERS EXACT TEST ARE'
        DO 20 I = 1, NUM
          WRITE (6,*) 'P(', I, ') = ', P(I)
   20   CONTINUE
        WRITE (6,*) 'PROBABILITY OF GIVEN TABLE IS IN P(',
     *    NPOS, ')'
      ELSE
        WRITE (6,*) 'EXPECTED FREQUENCIES ARE'
        DO 30 I = 1, NR
          WRITE (6,'(10F8.2)') (PRED(I,J),J=1,NC)
   30   CONTINUE
        WRITE (6,*) 'CHI-SQUARED STATISTIC = ', CHIS
        WRITE (6,*) 'NUMBER OF DEGREES OF FREEDOM = ', NDF
      END IF

      STOP

      END
```

G01AFF specimen run for Table (a)

```
ENTER THE NUMBER OF ROWS IN THE TABLE
 2
ENTER THE NUMBER OF COLUMNS IN THE TABLE
 3
ENTER TABLE FREQUENCIES ROW BY ROW
 28    60    57
 12    44    53
EXPECTED FREQUENCIES ARE
 22.83   59.37   62.80
 17.17   44.63   47.20
CHI-SQUARED STATISTIC = 3.984675
NUMBER OF DEGREES OF FREEDOM = 2
```

continued

G01AFF specimen run for Table (b)

```
ENTER THE NUMBER OF ROWS IN THE TABLE
 2
ENTER THE NUMBER OF COLUMNS IN THE TABLE
 2
ENTER TABLE FREQUENCIES ROW BY ROW
 11   2
  3   9
PROBABILITIES NEEDED FOR FISHERS EXACT TEST ARE
P(1)  = 1.7498990E-05
P(2)  = 7.6995558E-04
P(3)  = 1.0586889E-02
P(4)  = 6.3521336E-02
P(5)  = 0.1905640
P(6)  = 0.3049024
P(7)  = 0.2667896
P(8)  = 0.1270427
P(9)  = 3.1760668E-02
P(10) = 3.8497779E-03
P(11) = 1.9248889E-04
P(12) = 2.6921524E-06
PROBABILITY OF GIVEN TABLE IS IN P(10)
```

6.13 Multiple regression: G02CJF
[A difficult routine]

The **purpose** of this routine is to perform a multiple linear regression of a response variable y onto a set of m explanatory variables $(x_1, x_2, \ldots, x_m)$, i.e., to find estimates $\hat{\beta}_1, \hat{\beta}_2, \ldots, \hat{\beta}_m$ for the regression coefficients $\beta_1, \beta_2, \ldots, \beta_m$ in the model

$$y = \beta_1 x_1 + \beta_2 x_2 + \ldots + \beta_m x_m + \varepsilon.$$

Comment:
The routine assumes that you have n observations of y and $x_1, \ldots, x_m$, and that each value of y is related to the corresponding values of $x_1, \ldots, x_m$ by equations:

$$\begin{aligned}
y_1 &= \beta_1 x_{1,1} + \ldots + \beta_m x_{1,m} + \varepsilon_1 \\
y_2 &= \beta_1 x_{2,1} + \ldots + \beta_m x_{2,m} + \varepsilon_2 \\
&\quad\quad \cdot \quad \cdot \quad \cdot \\
y_n &= \beta_1 x_{n,1} + \ldots + \beta_m x_{n,m} + \varepsilon_n \, .
\end{aligned}$$
(6.9)

Here, y_i denotes the value of y in the i(th) observation, and $x_{i,j}$ denotes the value of x_j in the i(th) observation. ε_i denotes the 'error' or 'noise' in the i(th) observation, and it is assumed that each ε_i is randomly and independently distributed from a normal distribution with zero mean and constant variance.

The equations (6.9) can be written in matrix form:

$$y = X\beta + \varepsilon$$

Specimen problem

Given the 8 observations:

Observation	y	x_1	x_2
1	-1.4	3.2	2.7
2	-0.7	4.4	4.4
3	-0.1	4.2	4.4
4	0.1	3.0	3.7
5	0.5	3.7	4.5
6	0.9	3.2	3.6
7	1.2	3.9	4.6
8	1.6	2.0	3.1

(i) find estimates $\hat{\beta}_1$ and $\hat{\beta}_2$ for the regression coefficients β_1 and β_2 in the normal regression model

$$y = \beta_1 x_1 + \beta_2 x_2 + \varepsilon \ ;$$

(ii) estimate the standard errors and the covariance of the estimates $\hat{\beta}_1$ and $\hat{\beta}_2$.

Comment:

It is common for a regression model to include a constant term β_0 as in

$$y = \beta_0 + \beta_1 x_1 + \beta_2 x_2 + \varepsilon \ .$$

This routine can be used to fit a model with a constant term; how to do so will be explained in the **Postscript**.

The **method** uses a standard linear least-squares algorithm based on a QR-factorization of the matrix X.

The **routine name** with parameters is

G02CJF(X,IX,Y,IY,N,M,IR,BETA,IB,SIGSQ,C,IC,IPIV,WK1, WK2,IFAIL).

Description of parameters

Parameters which require values before G02CJF is called

N: [an integer variable]
N must specify the number of observations.

In the specimen problem, there are 8 observations, so N must be given the value 8.

M: [an integer variable]
M must specify the number of explanatory variables $x_1, \ldots, x_m$ in the regression model.

In the specimen problem, there are 2 explanatory variables x_1 and x_2, so M must be given the value 2.

Comment:
There must be at least as many observations as there are explanatory variables, otherwise there is not enough information in the data to allow a model to be fitted. So there is a restriction $M \leqslant N$.

IR: [an integer variable]
IR must specify the number of response variables.

This routine allows you to fit regression models for several response variables simultaneously, all against the same set of explanatory variables. However, in the interests of simplicity, the description of the routine given in this book is restricted to the common situation in which there is only one response variable. In this case IR must be given the value 1.

Details of how to handle several response variables simultaneously are given in the *NAG Manual*.

X: [a real two-dimensional array. It must have at least N rows and at least M columns in the REAL declaration in your calling program]

**Ref.
p. 19**
The i(th) row of X is used to contain the values of the explanatory variables $(x_1, \ldots, x_m)$ in the i(th) observation.

So, for the specimen problem

$$X(1,1) = 3.2 \quad X(1,2) = 2.7$$
$$X(2,1) = 4.4 \quad X(1,2) = 4.4$$
$$X(3,1) = 4.2 \quad X(3,2) = 4.4$$

and so on.

Y: [a real two-dimensional array. It must have at least N rows and at least IR columns in the REAL declaration in your calling program]
When there is only one response variable, the i(th) observation of y must be stored in $Y(I,1)$.

So for the specimen problem:

$$Y(1,1) = -1.4$$
$$Y(2,1) = -0.7$$
$$Y(3,1) = -0.1$$

and so on

$$\left.\begin{array}{l} \text{IX} \\ \text{IY} \\ \text{IC} \\ \text{IB} \end{array}\right\} :[\text{integer variables}]$$

Ref.
§2.4
IX, IY, IC, and IB must contain the size of the first dimensions of the arrays X, Y, C, and BETA respectively as they appear in the declaration in your calling program. The arrays C and BETA are described later.

So, if you had a (not very realistic) declaration

REAL X(5,4), Y(6,2), C(4,3), BETA(7,1) ,

in your calling program, then the statements

    IX = 5
    IY = 6
    IC = 4
and IB = 7

would also be needed.

Parameters associated with workspace

IPIV: [an integer one-dimensional array. Its length must be at least M in the INTEGER declaration in your calling program]
Ref.
§3.3
This array is used as workspace.

WK1: [a real two-dimensional array. It must have at least M rows and at least 4 columns in the REAL declaration in your calling program]
This array is used as workspace.

WK2: [a real one-dimensional array. Its length must be at least N in the REAL declaration in your calling program]
This array is used as workspace.

Thus, in the case of the specimen problem, where N = 8 and M = 2, minimum declarations

    REAL WK1(2,4), WK2(8)
    INTEGER IPIV(2)

are needed in your program.

In fact, it is advisable to declare more space than this, so that the program can also be used for larger problems.

The error parameter

IFAIL: [an integer variable]

IFAIL is the error parameter described in §3.2. It is recommended that you set

IFAIL = 0

before you call G02CJF. Then in the event of the routine failing, your program will stop and print one of the following error messages:

Error message	Meaning	Advice
IFAIL = 1	No solution can be obtained: the columns of X are linearly dependent.	Check your *x* variables. They are not mutually independent.
IFAIL = 2	No solution can be obtained: the matrix X is ill-conditioned.	Check your *x* variables. They show signs of mutual dependence.
IFAIL = 3	*Either* IX < N *or* IY < N *or* IC < N.	Check that you have declared the arrays X, Y, C, and BETA correctly, along with the corresponding values of IX, IY, IC, and IB.
IFAIL = 4	IB < M.	
IFAIL = 5	M > N.	M must be less than or equal to N.

If in any doubt as to what to do if an error message comes up, get some help.

Parameters to be examined after calling G02CJF

BETA: [a real two-dimensional array. It must have at least M rows and at least IR columns in the REAL declaration in your calling program]
The estimate of the j(th) regression coefficient will be found in BETA(J,1). Thus, for the specimen problem the estimates $\hat{\beta}_1$ and $\hat{\beta}_2$ will be found in BETA(1,1) and BETA(2,1).

SIGSQ: [a real one-dimensional array. Its length must be at least IR in the REAL declaration in your calling program]
SIGSQ(1) will contain an estimate of the variance of *y*.

Comment:
SIGSQ(1) is needed to compute the standard errors and covariances of the regression coefficient estimates.

C: [a real two-dimensional array. It must have at least N rows and at least M columns in the REAL declaration in your calling program]

The contents of this array are needed, together with SIGSQ(1), to compute the standard errors and covariances of the estimates of the regression coefficients. The estimated standard error of $\hat{\beta}_i$ is given by

SQRT(SIGSQ(1)*C(I,I)).

The estimated covariance between $\hat{\beta}_i$ and $\hat{\beta}_j$ $(i < j)$ is given by

SIGSQ(1)*C(I,J) .

So, for the specimen problem:

standard error of $\hat{\beta}_1$ = SQRT(SIGSQ(1)*C(1,1))
standard error of $\hat{\beta}_2$ = SQRT(SIGSQ(1)*C(2,2))
covar($\hat{\beta}_1$, $\hat{\beta}_2$) = SIGSQ(1)*C(1,2) .

Comments:

(i) Only elements C(I,J) with I < J must be used to compute covariances. This is no restriction since covar($\hat{\beta}_i$, $\hat{\beta}_j$) = covar($\hat{\beta}_j$, $\hat{\beta}_i$).

(ii) You might reasonably ask why the routine does not calculate the standard errors and covariances for you. The reason is that the routine can handle more than one response variable, and then a different array C would be needed to store the standard errors and covariances of the estimates for each response variable. For an exact description of the contents of C, consult the *NAG Manual*.

Specimen program

Program planning

1. *Declare*   REAL X(,), Y(,1), BETA(,1), SIGSQ(1), C(,), WK1(,4), WK2( ), SE, COVAR
   INTEGER IX, IY, N, M, IR, IB, IC, IPIV( ), IFAIL

2. *Set*   IX, IY, IB, IC, IR
   Read   N, M, X, Y
   Set   IFAIL = 0

3. *Call*   G02CJF

4. *Calculate* the standard errors and covariances of the $\hat{\beta}_i$

5. *Print*   BETA and the standard errors and covariances.

Comments:

(i) The following program makes allowance for:

(a) up to 10 explanatory variables $x_1, \ldots, x_{10}$.
(b) up to 20 observations.

If your problem involves quantities larger than this, then you will have to alter the dimensions of the arrays in the declaration in the calling program

**Ref.
§5.2**
accordingly. You will also have to change the values given to IX, IY, IB, and IC in the program. You will also find it convenient to read in the values of X and Y from a data file.

(ii) You may need to make changes to this program in order to make it run correctly on your computer. *See §3.4 for details.*

G02CJF specimen program

```
C       G02CJF: MULTIPLE REGRESSION

        REAL X(20,10), Y(20,1), BETA(10,1), SIGSQ(1), C(20,10)
        REAL WK1(10,4), WK2(20), SE, COVAR
        INTEGER IX, IY, N, M, IR, IB, IC, IPIV(10), IFAIL, I,
     *          J

        IX = 20
        IY = 20
        IR = 1
        IB = 10
        IC = 20

        WRITE (6,*)
     *  'ENTER THE NUMBER OF EXPLANATORY VARIABLES'
        READ (5,*) M
        WRITE (6,*) 'ENTER THE NUMBER OF OBSERVATIONS'
        READ (5,*) N
        WRITE (6,*)
     *  'ENTER THE Y AND X VALUES FOR EACH OBSERVATION'
        WRITE (6,*) 'FOR EACH OBSERVATION START A NEW LINE'
        DO 10 I = 1, N
           READ (5,*) Y(I,1), (X(I,J),J=1,M)
     10 CONTINUE

        IFAIL = 0

        CALL G02CJF(X,IX,Y,IY,N,M,IR,BETA,IB,SIGSQ,C,IC,IPIV,
     *              WK1,WK2,IFAIL)

        DO 20 I = 1, M
           SE = SQRT(SIGSQ(1)*C(I,I))
           WRITE (6,*) 'ESTIMATE FOR BETA(', I, ') IS',
     *       BETA(I,1)
           WRITE (6,*) '   WITH STANDARD ERROR', SE
     20 CONTINUE
        DO 40 I = 1, M
           DO 30 J = I + 1, M
              COVAR = SIGSQ(1)*C(I,J)
              WRITE (6,*) 'COVARIANCE ESTIMATE FOR BETA(', I,
     *          ') AND BETA(', J, ') IS', COVAR
     30    CONTINUE
     40 CONTINUE

        STOP

        END
```

G02CJF specimen run

```
ENTER THE NUMBER OF EXPLANATORY VARIABLES
2
ENTER THE NUMBER OF OBSERVATIONS
8
ENTER THE Y AND X VALUES FOR EACH OBSERVATION
 FOR EACH OBSERVATION START A NEW LINE
-1.4   3.2   2.7
-0.7   4.4   4.4
-0.1   4.2   4.4
 0.1   3.0   3.7
 0.5   3.7   4.5
 0.9   3.2   3.6
 1.2   3.9   4.6
 1.6   2.0   3.1
ESTIMATE FOR BETA(1) IS   -1.780288
        WITH STANDARD ERROR   0.3589033
ESTIMATE FOR BETA(2) IS   1.654882
        WITH STANDARD ERROR   0.3217193
COVARIANCE ESTIMATE FOR BETA(1) AND BETA(2) IS   -0.1144494
```

Postscript

Comment:

To use this routine to fit a regression model with a *constant term*, you must simply define an extra explanatory variable whose value in each observation is 1.0. The estimated regression coefficient for this extra variable is then the estimate for the constant term.

6.14 Number sort (descending order): M01AKF [An easy routine]

The **purpose** of this routine is to sort a given set of numbers into descending order.

Specimen problem

To sort the 7 numbers

$$2.7 \quad -3.8 \quad 5.1 \quad 4.3 \quad 3.8 \quad 6.2 \quad 1.2$$

into descending order.

The **method** used is a recursive list-merge algorithm.

The **routine name** with parameters is

M01AKF (A, W, IND, INDW, N, NW, IFAIL).

Description of parameters

Parameters which require values before M01AKF is called

N: [an integer variable]
N is the number of numbers to be sorted in the data list.

In the specimen problem, there are 7 numbers. So, N would be given the value 7 in this instance.

A: [a real one-dimensional array. Its length must be at least N in the REAL declaration in your calling program]

Ref.
p. 19   The array A holds the numbers to be sorted.

So, in the case of the specimen problem

$$A(1) = 2.7, \quad A(2) = -3.8, \quad \text{and so on.}$$

Parameters associated with workspace

W $\left. \right\}$ [respectively, a real one-dimensional array and an integer one-
INDW $\left. \right\}$ dimensional array. Both arrays must have length at least 1 in the

Ref.
§3.3   declarations in your calling program]
Both these arrays are used as workspace.

Comments:

(i) Both these arrays should have the same length in the declarations in your calling program.

(ii) The more workspace you allow, the quicker the routine sorts the numbers. A good compromise for an appropriate amount of workspace is about a tenth of the size of the numbers which your program will sort. Thus, if your program had a declaration

REAL A(1000),

Then suitable workspace declarations would be

REAL W(100)
and   INTEGER INDW(100).

NW: [an integer variable]
NW must specify the length of the array W (and also of INDW) as declared in your calling program.

So, if your program had the declarations above, then a statement

NW = 100

would also be needed.

The error parameter

IFAIL: [an integer variable]

IFAIL is the error parameter described in §3.2. It is recommended that you set

$$IFAIL = 0$$

before you call M01AKF. Then in the event of the routine failing, your program will stop and print one of the following messages:

Error message	Meaning	Advice
IFAIL = 1	N ≤ 0.	
IFAIL = 2	NW ≤ 0.	Check that you gave positive (correct) values to N and NW.
IFAIL = 3	N ≤ 0 and NW ≤ 0.	

Parameters to be examined after calling M01AKF

A: [the one-dimensional array described earlier]

Before M01AKF is called the array A is used to contain the unsorted numbers. However, after a successful call of the routine, A(1), A(2), . . ., A(N) will contain the numbers in descending order.

IND: [an integer one-dimensional array. Its length must be at least N in the INTEGER declaration in your calling program]

IND(K) will tell you the position that the k(th) sorted number had in the original list.

For instance, looking at the data in the specimen problem, A(1) will contain 6.2 after a successful call of the routine, and IND(1) will contain 6. This tells us that 6.2 was the 6(th) number on the original list.

Specimen program

Program planning

1. *Declare* REAL A( ), W( )
          INTEGER IND( ), INDW( ), N, NW, IFAIL

2. *Set*   NW
   Read   N, A
   Set   IFAIL

3. *Call*   M01AKF

4. *Print* A, IND

Comments:

(i) The following program can be used to sort up to 1000 numbers into descending order. If you have more numbers than this, you will have to alter the dimensions in the declarations in the program accordingly. You will also have to change the value given to NW in the program.

(ii) You may need to make some changes to this program in order to make it run correctly on your computer. *See §3.4 for details.*

M01AKF specimen program

```
C       M01AKF: NUMBER SORT (DESCENDING ORDER)

        REAL A(1000), W(100)
        INTEGER IND(1000), INDW(100), N, NW, IFAIL, I

        NW = 100

        WRITE (6,*) 'ENTER THE NUMBER OF NUMBERS TO BE SORTED'
        READ (5,*) N
        WRITE (6,*) 'ENTER THE NUMBERS TO BE SORTED'
        READ (5,*) (A(I),I=1,N)

        IFAIL = 0

        CALL M01AKF(A,W,IND,INDW,N,NW,IFAIL)

        WRITE (6,*)
     *   'SORTED NUMBERS      PLACE IN ORIGINAL ARRAY'
        DO 10 I = 1, N
           WRITE (6,*) A(I), '                  ', IND(I)
     10 CONTINUE

        STOP

        END
```

M01AKF specimen run

```
ENTER THE NUMBER OF NUMBERS TO BE SORTED
 7
ENTER THE NUMBERS TO BE SORTED
 2.7  -3.8   5.1    4.3    3.8    6.2    1.2
SORTED NUMBERS      PLACE IN ORIGINAL ARRAY
6.200000             6
5.100000             3
4.300000             4
3.800000             5
2.700000             1
1.200000             7
-3.800000            2
```

Postscript

Comment:

Often, the knowledge of where a number was in the original list is important. With each number in the original array, there might be another parallel array holding an associated piece of information (such as a name). After the numbers have been sorted, the values in IND would enable you to print out the names alongside the corresponding numbers.

Associated routines:

(i) If you want to sort a set of numbers into ascending order, then you could use M01AJF. The description for M01AJF is identical to that for M01AKF except for the order in which the numbers are printed. So you could adapt the description for M01AKF given here.

(ii) If you just want to sort a set of numbers without at the same time generating an index array IND, then the routines M01APF (descending order) or M01ANF (ascending order) are faster than M01AKF or M01AJF. You will find details of these routines in the *NAG Manual*.

(iii) The sorting routines in the *NAG Library* are only suitable for sorting data which can be stored in an array. If you wish to sort a larger file of data, use some other sorting package, rather than a NAG routine.

6.15 Evaluation of Bessel function $J_0(x)$: S17AEF
Evaluation of Bessel function $J_1(x)$: S17AFF
[Easy routines]

Comments:

(i) These routines differ from other routines described in this book. They are both set up as functions, rather than as subroutines. In fact, all the routines in this chapter of the *NAG Manual* are set up as functions. There are a sprinkling of other routines in the *NAG Manual* which are set up in this way (notably some of the integration routines). So it is a good idea to keep a watchful eye on whether a NAG routine is specified as a function or as a subroutine.

(ii) The description of these routines is so similar, that with little amendment the same description suffices for both routines.

The **purpose** of S17AEF and S17AFF is to evaluate the Bessel functions $J_0(x)$ and $J_1(x)$ respectively for a given value of x.

Specimen problem

To find the values of $J_0(x)$ and $J_1(x)$ for $x = 0.5$.

The **method** of approximation uses Chebyshev polynomials.

The **routine names** with parameters are:

S17AEF (X, IFAIL)
S17AFF (X, IFAIL).

Comment:
These routines are functions, not subroutines. This means that once X and IFAIL have been specified, the statement

Y = S17AEF (X, IFAIL)

has the effect of giving Y the value $J_0(X)$.

Description of parameters

Both parameters require values before either S17AEF or S17AFF are used.

X: [a real variable]
X must contain the number x whose Bessel function $J_0(x)$ or $J_1(x)$ is required. So, in the case of the specimen problem X must be set to 0.5.

IFAIL: [an integer variable]
IFAIL is the error parameter described in §3.2. It is recommended that you set

IFAIL = 0

before you use S17AEF or S17AFF. Then in the event of either routine failing, your program will stop and print

IFAIL = 1.

This means that the value specified for X was too large for the routine to give an answer with any accuracy.

Comment:
It is not possible to indicate exactly the largest value of x for which the routine can evaluate the Bessel function. This value depends on the particular machine which you are using.

Specimen program

Program planning

1. *Declare* REAL X, J0, J1, S17AEF, S17AFF
   INTEGER IFAIL

2. *Read* X
   Set IFAIL

3.   J0 = S17AEF
     J1 = S17AFF

4. *Print* J0, J1

Comment:

You may need to make some changes to this program in order to make it run correctly on your computer. *See §3.4 for details.*

S17AEF/S17AFF specimen program

```
C       S17AEF: BESSEL FUNCTION J0
C       S17AFF: BESSEL FUNCTION J1

        REAL X, J0, J1, S17AEF, S17AFF
        INTEGER IFAIL

        WRITE (6,*) 'ENTER A VALUE X'
        READ (5,*) X

        IFAIL = 0

        J0 = S17AEF(X,IFAIL)
        J1 = S17AFF(X,IFAIL)

        WRITE (6,*) 'VALUE FOR J0(', X, ') IS ', J0
        WRITE (6,*) 'VALUE FOR J1(', X, ') IS ', J1

        STOP

        END
```

S17AEF/S17AFF specimen run

```
ENTER A VALUE X
 0.5
VALUE FOR J0(0.5000000) IS 0.9384698
VALUE FOR J1(0.5000000) IS 0.2422685
```

7
The NAG Graphical Supplement

7.1 J06: Introduction

The *NAG Graphical Supplement* consists of a series of routines which allow you to display data and results in pictorial or graphical form. There are routines for drawing and plotting curves, drawing contour maps and surfaces, and so on.

The routines selected for this book are intended to familiarize you with some of the ideas of the Supplement, and to provide you with some easy routines with which to work. However, before you read any further, *you ought to check that the NAG graphical routines are currently available on your particular machine*.

Many computer installations have other sets of graphical routines available. For instance, if your institution has a CALCOMP graph-plotter, then you may well have access to the CALCOMP routines. A single NAG routine might well use four or five of these underlying routines to do its job. So one of the advantages of using the NAG routines is that most of the detailed work of sending instructions to a graph-plotter or terminal is done for you.

As well as enabling you to produce graphs using a graph-plotter, the *NAG Graphical Supplement* allows you to produce rough pictures on an ordinary terminal or a lineprinter. Thus you can see if any adjustment is needed to your graph before you ask for it to be plotted on a graphics terminal or on a graph-plotter. This can be very useful because getting a graph plotted on a graph-plotter may involve a considerable delay (and possibly some expense). The other advantage of using the NAG routines is that you can put a program together which is largely machine independent. The resulting program could, with very little change, be used on any computer which has the NAG graphical routines.

Before you run a program which uses a NAG graphical routine, you will have to find out what instructions are needed to access the NAG graphical routines on your computer. These can be quite different for each machine, and different from the instructions needed to access the numerical routines. So, *check what to do at your computer centre.*

7.2 A basic framework

Use of the *NAG Graphical Supplement* requires a bit more initial effort than using the NAG numerical routines. However, once a basic framework has been established, it is just as straightforward.

The output from a NAG graphical routine depends on a number of choices such as the type of plotter, the shape and size of the required plot, the colour of ink to be used, the style of lines to be drawn, and so on. All these things have to be specified in a program from which a graph is required. NAG provides a number of simple routines for setting these various options, some of which have to be called from any program which includes graphical output.

This section describes a minimum set of these option setting routines. The descriptions are intentionally brief. If you need more information about these, consult the appropriate *NAG Manual*. (Routines whose names begin with J06 are described in a separate *NAG Manual* called the '*Graphical Supplement*'.)

The order in which these basic routines should be called is summarised in §7.2.4. The framework given there is used in all the specimen programs in this chapter.

7.2.1 Destination of graphical output

The most important information that you have to specify is on what piece of equipment or device you want your graph to appear. How this is specified varies from one computer to another.

It is recommended that, when you run a program which produces a graph, you first look at a rough picture on an ordinary terminal or lineprinter – to see, for example, whether the graph is in the right position on the page. This is done by a call to the NAG routine X04ABF. This call must precede any call to a NAG graphical-plotting routine. A statement

CALL X04ABF (1, 6)

ensures that the rough picture is sent to device number 6. In the specimen programs in this book, it is assumed that this device number corresponds to an ordinary terminal.

When you are satisfied with the rough picture you have obtained, you will want to instruct the computer to produce the graph on a graphics terminal or a graph-plotter. To do this, the call to X04ABF must be replaced by a call to some other routine. Which routine to use will depend on what computer you are using and on the graphical devices which are available. In the specimen programs in this book, the call to a CALCOMP routine

CALL PLOTS (0, 0, 5)

is used as an example. You will have to check at your computer centre to find exactly what routine to call, and how to set the parameters. (The purpose of this routine is to initialize the software which drives the graphical device that you want to use.)

Comment:
You *must* call PLOTS (or an equivalent routine) or X04ABF, before you call any of the NAG J06 routines.

Important advice:
When you are ready to run a program which includes calls to NAG graphical routines, you will have to find out the appropriate instruction (or instructions) for doing this at your computer centre. The instruction needed to run a program which calls 'ordinary' NAG routines will differ from the one needed to run a program which produces a graph, which again will be different from the one needed to run a program which gives a rough picture. So, it is important in each case that you check what the appropriate instruction is at your computer centre.

7.2.2 Initial settings: J06WAF, J06WBF

Having called X04ABF or PLOTS (or an equivalent routine), you must next call a NAG routine which sets various options for the graphs which you want to draw. The two routines described below are concerned with setting these options.

J06WAF

The **purpose** of this routine is to make various automatic initial choices or option settings.

A routine call

CALL J06WAF

causes the following options to be set automatically:
   (i) The graph will occupy the largest square which can be fitted onto the plotting surface (i.e. terminal screen or page of paper).
   (ii) The (x,y) co-ordinates of the corners of the region to be plotted will be set to $(0,0)$, $(1,0)$, $(1,1)$, $(0,1)$.
   (iii) Other initial settings are chosen, such as what colour pens to use, how thick the line should be, how smoothly curved lines should be drawn, and so on.

Comment:
This routine has no parameters, as everything is done automatically. Even if you don't like the initial settings, *this routine must be called*, as it

starts off the whole NAG graphical procedure. You will probably be quite happy with most of these settings. However, you can, by calling other routines, override the choices made by J06WAF. In particular, it is unlikely that your chosen graph will always lie in the unit square. The following routine overrides the automatic choice of co-ordinates made in (ii) above.

J06WBF

The **purpose** of this routine is to reset the co-ordinates of the region to be plotted.

Comment:
Normally, you will want to choose the region to be plotted to coincide with the range of values of x and y in your graph.

The **routine name** with parameters is

J06WBF (XMIN, XMAX, YMIN, YMAX, MARGIN).

Description of parameters

Note:
All the parameters in this routine require values before the routine is called.

XMIN ⎫
XMAX ⎭ : [real variables]
XMIN and XMAX should normally contain the minimum and maximum values of x for which you want to plot the graph.

YMIN ⎫
YMAX ⎭ : [real variables]
YMIN and YMAX should normally contain the minimum and maximum values of y for which you want to plot the graph.

Comment:
In some cases, you may not know the minimum and maximum values of y exactly. In this event, make a guess at them, have a look at a rough picture using X04ABF. The picture should help you to choose better values for YMIN and YMAX if necessary.

MARGIN: [an integer variable]
MARGIN must be set to either 0 or 1.

Value given to MARGIN	Result
MARGIN = 0	No margin.
MARGIN = 1	A margin is placed all round your graph.

Comment:
The following diagram shows how the region to be plotted (defined by your (x,y) co-ordinates) is mapped onto the plotting surface.

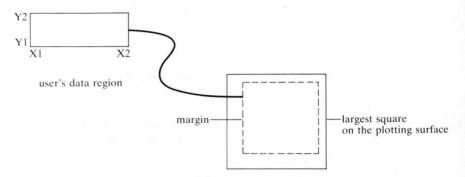

user's data region

margin—

—largest square
on the plotting surface

(*Note*: Figure taken from Essential Introduction to the *NAG Graphical Supplement*.)

7.2.3 Finishing off: J06WZF

At the end of a program which calls graphical routines, you must call the routine J06WZF.

J06WZF

The **purpose** of this routine is to finish off the graphical procedure, and ensure that the plot is actually produced.

Note:
This routine has no parameters.

7.2.4 Summary

In brief, any program from which you want to obtain some graphical output using the *NAG Graphical Supplement* will have to contain calls to the following routines specified in the order given opposite.

The five basic routines described in the following sections will all fit into this framework.
Finally, you will observe in the following descriptions that there are seldom any parameters to be printed after a routine has been called. This is because the graphical routines produce the graphs automatically.

```
      MARGIN = 1

      WRITE (6,*) 'ENTER THE LEAST AND GREATEST VALUES OF X'
      READ (5,*) XMIN, XMAX
      WRITE (6,*) 'ENTER THE LEAST AND GREATEST VALUES OF Y'
      READ (5,*) YMIN, YMAX

      CALL PLOTS(0,0,5)        (or equivalent routine)
      CALL J06WAF
      CALL J06WBF(XMIN,XMAX,YMIN,YMAX,MARGIN)

          ┌─────────────────────────────────────────┐
          │ The instructions for the particular job   │
          │    you want to do will go in here.        │
          └─────────────────────────────────────────┘

      CALL J06WZF

      STOP

      END
```

The following descriptions and specimen programs should give you enough experience to tackle the other routines in the *NAG Graphical Supplement* with confidence.

7.3 Axes: J06AAF
[A very easy routine]

The **purpose** of this routine is to draw a pair of axes.

> *Comment*:
> This routine has no parameters. It picks up and uses the values which you gave to XMIN, XMAX, YMIN, and YMAX in J06WBF.

The **routine name** is

J06AAF.

Specimen program

Problem

To put in axes on a graph so that the horizontal x-axis goes between -2 and 5, and the vertical y-axis goes between -1 and 2.

Program planning

    1. *Declare* REAL XMIN, XMAX, YMIN, YMAX
                    INTEGER MARGIN

    2. *Use*   the framework in §7.2.4.

    3. *Insert*:
        Call   J06AAF

Comments:

    (i) The following very simple program is included to demonstrate the framework in §7.2.4 in action.

    (ii) You may need to make some changes to this program in order to make it run correctly on your computer. *See §3.4 for details.* You may also need to replace the statement

        CALL PLOTS (0, 0, 5)

by one appropriate to your computer.

J06AAF specimen program

```
C      J06AAF: PLOTTING AXES

       REAL XMIN, XMAX, YMIN, YMAX
       INTEGER MARGIN

       MARGIN = 1

       WRITE (6,*) 'ENTER THE LEAST AND GREATEST VALUES OF X'
       READ (5,*) XMIN, XMAX
       WRITE (6,*) 'ENTER THE LEAST AND GREATEST VALUES OF Y'
       READ (5,*) YMIN, YMAX

       CALL PLOTS(0,0,5)
       CALL J06WAF
       CALL J06WBF(XMIN,XMAX,YMIN,YMAX,MARGIN)

      ┌─────────────────────────┐
      │   CALL J06AAF           │
      └─────────────────────────┘

       CALL J06WZF

       STOP

       END
```

J06AAF specimen run

(a) *on the terminal*

```
ENTER THE LEAST AND GREATEST VALUES OF X
-2.0    5.0
ENTER THE LEAST AND GREATEST VALUES OF Y
-1.0    2.0
```

(b) *on the graph-plotter*

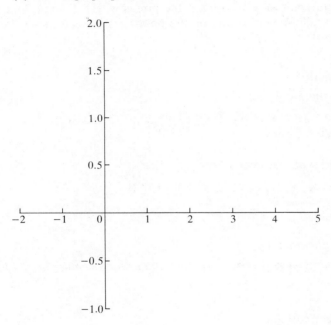

Postscript

Associated routines:

(i) After using J06AAF, if you want to label the axes, use J06AJF.

(ii) If you want to give a title to a graph, use J06AHF.

You will find details of these routines in the *NAG Graphical Supplement*.

7.4 Plotting points: J06BAF
[An easy routine]

The **purpose** of this routine is to plot a given set of data points (x_1,y_1), (x_2,y_2), . . ., (x_n,y_n).

Comment:
The routine allows you to mark the points with a symbol, and also, if required, allows you to join up the points with straight lines.

Specimen problem

To plot the 5 points

(2.5,18.75), (5.0,4.46), (10.0,1.502), (15.0,0.832) and (20.0,0.566).

The **routine name** with parameters is

J06BAF (X, Y, N, ITYPE, KSYM, IFAIL).

Description of parameters

All the parameters described in this routine require values before J06BAF is called.

N: [an integer variable]
N should contain the number of points which you want to plot.

In the case of the specimen problem, there are 5 points. So, in this case, N should be given the value 5.

Comment:
This routine needs at least 1 point to plot (i.e. $N \geq 1$).

$\left.\begin{array}{c} X \\ Y \end{array}\right\}$: [real one-dimensional arrays. Both arrays must have length at least N in the REAL declaration in your calling program]

Ref.
p. 19

X(K) and Y(K) are used to contain the x- and y-values respectively of the k(th) data point (x_k,y_k).

So, in the case of the specimen problem, you would specify

$$X(1) = 2.5 \quad Y(1) = 18.75$$
$$X(2) = 5.0 \quad Y(2) = 4.46$$

and so on.

KSYM: [an integer variable]

KSYM must be given an integer value between 1 and 9. The value which you give to KSYM will determine what symbol is printed at each data point.

The range of possible symbols normally available is given in the following table:

Value of KSYM	1	2	3	4	5	6	7	8	9
Symbol produced	×	⊙	+	∗	⊡	◇	▽	△	+

Thus, if you wanted the symbol '×' to be printed at each plotted point, you would specify KSYM = 1.

ITYPE: [an integer variable]

ITYPE must be given an integer value between −2 and 2. The value given to ITYPE controls whether just the points are plotted, whether lines joining the points are put in, or whether the lines are joined up to form a closed polygon.

| Value given to ITYPE | Effect: | | |
	Symbols plotted	Lines joining the points	Closed polygon
ITYPE = 2	√	√	
ITYPE = 1		√	
ITYPE = 0	√		
ITYPE = −1		√	√
ITYPE = −2	√	√	√

Thus, if you wanted the points plotted only, you would specify ITYPE = 0.

Comment:

If ITYPE is given a value outside the values specified above, then the routine automatically assumes that ITYPE = 1.

The error parameter

>IFAIL: [an integer variable]
>
>>IFAIL is the error parameter described in §3.2. It is recommended that you set
>>
>>>IFAIL = 0
>>
>>before you call J06BAF. Then in the event of the routine failing, your program will stop and print
>>
>>>IFAIL = 1.
>>
>>This occurs only in the event that no data points have been specified (i.e. N = 0).

Specimen program

Program planning

>1. *Declare* REAL X( ), Y( ), XMIN, XMAX, YMIN, YMAX
>   INTEGER N, ITYPE, KSYM, IFAIL, MARGIN
>
>2. *Use* the framework in §7.2.4.
>
>>3. *Insert*
>>>Call J06AAF (Axes)
>>>Set   KYSM, ITYPE
>>>Read N, X, Y
>>>Set   IFAIL
>>>Call J06BAF

>*Comments*:
>
>   (i) The following program can be used to plot up to 30 points. If you want to use the program for more points, then you will have to alter the lengths of the arrays X and Y in the REAL declaration in the program accordingly.
>
>   (ii) You may need to make some changes to this program in order to make it run correctly on your computer. *See §3.4 for details*. You may also need to replace the statement
>
>>CALL PLOTS (0, 0, 5)
>
>by one appropriate to your computer.

J06BAF specimen program

```
C        J06BAF: PLOTTING POINTS
C        J06AAF: (WITH AXES)

         REAL X(30), Y(30), XMIN, XMAX, YMIN, YMAX
         INTEGER N, ITYPE, KSYM, IFAIL, MARGIN, I

         MARGIN = 1

         WRITE (6,*) 'ENTER THE LEAST AND GREATEST VALUES OF X'
         READ (5,*) XMIN, XMAX
         WRITE (6,*) 'ENTER THE LEAST AND GREATEST VALUES OF Y'
         READ (5,*) YMIN, YMAX

         CALL PLOTS(0,0,5)
         CALL J06WAF
         CALL J06WBF(XMIN,XMAX,YMIN,YMAX,MARGIN)
```

```
         CALL J06AAF

         ITYPE = 0
         KSYM = 1

         WRITE (6,*) 'ENTER THE NUMBER OF POINTS'
         READ (5,*) N
         WRITE (6,*)
       *  'ENTER VALUES OF X AND Y, ONE PAIR PER LINE'
         DO 10 I = 1, N
           READ (5,*) X(I), Y(I)
      10 CONTINUE

         IFAIL = 0

         CALL J06BAF(X,Y,N,ITYPE,KSYM,IFAIL)
```

```
         CALL J06WZF

         STOP

         END
```

J06BAF specimen run

(a) *on the terminal*

```
ENTER THE LEAST AND GREATEST VALUES OF X
 0.0    20.0
ENTER THE LEAST AND GREATEST VALUES OF Y
 0.0    20.0
ENTER NUMBER OF POINTS
 5
ENTER VALUES OF X AND Y, ONE PAIR PER LINE
 2.5    18.75
 5.0     4.46
10.0     1.502
15.0     0.832
20.0     0.566
```

continued

(b) *on the graph-plotter*

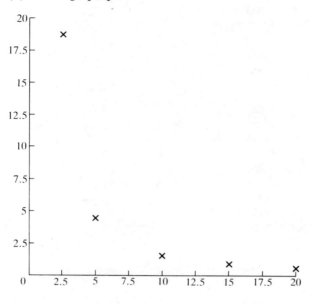

Postscript

Associated routines:
If you want a smooth curve through these points, use this routine in conjunction with J06CAF, which is described in §7.5.

7.5 Smooth curve through points: J06CAF [An easy routine]

The **purpose** of this routine is to draw a smooth curve through a given set of data points (x_1, y_1), (x_2, y_2), . . ., (x_n, y_n).

Specimen problem

To put a smooth curve through the 5 points

(2.5,18.75), (5.0,4.46), (10.0,1.502), (15.0,0.832) and (20.0,0.566).

The **routine name** with parameters is

J06CAF (X, Y, N, METHOD, IFAIL).

Description of parameters

All the parameters described in this routine require values before J06CAF is called.

N: [an integer variable]
N should contain the number of points through which you want the curve to pass.

In the case of the specimen problem above there are 5 points, so in this case N should be given the value 5.

Comment:
A curve needs to pass through at least 2 points, so N must be at least 2.

X⎫ [real one-dimensional arrays. Both arrays must have length at least N in
Y⎭ the REAL declaration in your calling program]

Ref.
p. 19
X(K) and Y(K) are used to contain the k(th) data point (x_k, y_k) through which the cuve is required to pass.

So, in the case of the specimen problem, you would specify

$$X(1) = 2.5 \quad Y(1) = 18.75$$
$$X(2) = 5.0 \quad Y(2) = \ \ 4.46$$

and so on.

Comment:
The data points must be arranged in an order so that the x-values are either non-decreasing

i.e. $X(1) \leqslant X(2) \leqslant \ldots \leqslant X(N)$.

or non-increasing

i.e. $X(1) \geqslant X(2) \geqslant \ldots \geqslant X(N)$.

METHOD: [an integer variable]
METHOD must be given the value either 1 or 2. The value given to METHOD controls which method is used to draw the curve.

Value given to METHOD	Method used
METHOD = 1	Piecewise monotonic method.
METHOD = 2	Cubic Bessel method.

Comment:
METHOD = 1 produces a much tighter-fitting curve than METHOD =2. You are advised to try METHOD = 2 first to see if it produces a satisfactory curve.

The error parameter

IFAIL: [an integer variable]

IFAIL is the error parameter described in §3.2. It is recommended that you set

IFAIL = 0

before you call J06CAF. Then in the event of the routine failing, your program will stop and print one of the following error messages:

Error message	Meaning	Advice
IFAIL = 1	*Either* N < 2 *or* METHOD ≠ 1 *or* 2	Check that you have given correct values to N and METHOD.
IFAIL = 2	The *x*-values in your data are incorrectly ordered.	Check that the data are ordered so that the *x*-values are either non-increasing or non-decreasing.

Specimen program

Program planning

1. *Declare* REAL X( ), Y( ), XMIN, XMAX, YMIN, YMAX
   INTEGER N, METHOD, IFAIL, MARGIN

2. *Use* the framework in §7.2.4.

3. *Insert*

   Call   J06AAF (Axes)
   Set   METHOD
   Read   N, X, Y
   Set   IFAIL = 0
   Call   J06CAF

Comments:

(i) The following program obtains a rough sketch on a terminal of a curve through a given set of points. If you want to use the program for more than 30 points, then you will have to alter the lengths of the arrays X and Y in the REAL declaration in the program accordingly.

(ii) You will have to choose the values for XMIN, XMAX, YMIN, and YMAX by looking at your data. In the case of the specimen problem you could chose XMIN = 0.0, XMAX = 20.0, YMIN = 0.0, and YMAX = 20.0.

(iii) If you are satisfied with the graph produced on the terminal, then just replace the call to X04ABF by a call to PLOTS (or equivalent) and run the program again, using the instruction needed to obtain a graph from the graph-plotter.

(iv) You may need to make some changes to this program in order to make it run correctly on your computer. *See §3.4 for details.* You may also need to replace the statement

CALL PLOTS (0, 0, 5)

by one appropriate to your computer.

J06CAF specimen programs
  (a) *Specimen program I*

Comment:
The first program obtains a rough picture of the graph on the terminal.

```
C       J06CAF: CURVE THROUGH POINTS (UNMARKED)
C       J06AAF: (WITH AXES)

        REAL X(30), Y(30), XMIN, XMAX, YMIN, YMAX
        INTEGER N, METHOD, IFAIL, MARGIN, I

        MARGIN = 1

        WRITE (6,*) 'ENTER THE LEAST AND GREATEST VALUES OF X'
        READ (5,*) XMIN, XMAX
        WRITE (6,*) 'ENTER THE LEAST AND GREATEST VALUES OF Y'
        READ (5,*) YMIN, YMAX

        CALL X04ABF(1,6)
        CALL J06WAF
        CALL J06WBF(XMIN,XMAX,YMIN,YMAX,MARGIN)
```

```
        CALL J06AAF

        METHOD = 2

        WRITE (6,*) 'ENTER NUMBER OF POINTS'
        READ (5,*) N
        WRITE (6,*)
     *   'ENTER VALUES OF X AND Y, ONE PAIR PER LINE'
        DO 10 I = 1, N
          READ (5,*) X(I), Y(I)
     10 CONTINUE

        IFAIL = 0

        CALL J06CAF(X,Y,N,METHOD,IFAIL)
```

```
        CALL J06WZF

        STOP

        END
```

continued

Specimen run I

```
ENTER THE LEAST AND GREATEST VALUES OF X
 0.0  20.0
ENTER THE LEAST AND GREATEST VALUES OF Y
 0.0  20.0
ENTER NUMBER OF POINTS
 5
ENTER VALUES OF X AND Y, ONE PAIR PER LINE
 2.5  18.75
 5.0   4.46
10.0   1.502
15.0   0.832
20.0   0.566
```

```
  20..
     .
     .       .
     .       .
     ..
17.5.       .
     .
     .        .
     .        .
 15..
     .        .
     .        .
     .        .
     ..        .
12.5.        .
     .
     .         .
     .         .
 10..         .
     .
     .          .
     .          .
7.5..          .
     .
     .           .
     .           .
     .            .
 5..            .
     .
     .            .
     .             .
     .              .
2.5..
     .               .
     .                 .   ............
     .                   ......            .......................
     .................................................................
  0
  0     2.5     5     7.5    10     12.5    15     17.5    20
```

A practised eye can see that this is going to produce the rather unsatisfactory graph that follows.

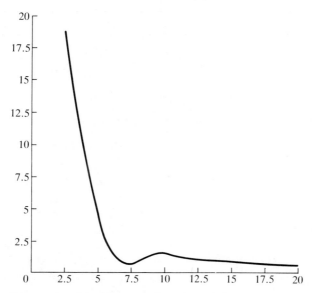

The marked bulge at the bottom of the curve does not really reflect the trend of the points, and also the points are not marked.

Comment:

Investigation of the results above suggests the following improvements:

   (i) Change the value given to METHOD. If METHOD is given the value 1, a tighter-fitting curve is obtained.

   (ii) A call of the routine J06BAF will produce the points marked on the curve.

The following specimen program implements these changes, and produces a graph on the graph-plotter.

(b) *Specimen program II*

```
C       J06BAF: PLOTTING POINTS
C       J06CAF: CURVE THROUGH POINTS
C       J06AAF: (WITH AXES)

        REAL X(30), Y(30), XMIN, XMAX, YMIN, YMAX
        INTEGER N, METHOD, IFAIL, ITYPE, KSYM, MARGIN, I

        MARGIN = 1

        WRITE (6,*) 'ENTER THE LEAST AND GREATEST VALUES OF X'
        READ (5,*) XMIN, XMAX
        WRITE (6,*) 'ENTER THE LEAST AND GREATEST VALUES OF Y'
        READ (5,*) YMIN, YMAX

        CALL PLOTS(0,0,5)
        CALL J06WAF
        CALL J06WBF(XMIN,XMAX,YMIN,YMAX,MARGIN)
```

continued

```
      CALL J06AAF

      ITYPE = 0
      KSYM = 1

      WRITE (6,*) 'ENTER NUMBER OF POINTS'
      READ (5,*) N
      WRITE (6,*)
    *    'ENTER VALUES OF X AND Y, ONE PAIR PER LINE'
      DO 10 I = 1, N
        READ (5,*) X(I), Y(I)
  10 CONTINUE

      IFAIL = 0

      CALL J06BAF(X,Y,N,ITYPE,KSYM,IFAIL)

      METHOD = 1


      CALL J06CAF(X,Y,N,METHOD,IFAIL)
```

```
      CALL J06WZF

      STOP

      END
```

Specimen run II
(a) *on the terminal*

```
ENTER THE LEAST AND GREATEST VALUES OF X
 0.0   20.0
ENTER THE LEAST AND GREATEST VALUES OF Y
 0.0   20.0
ENTER NUMBER OF POINTS
 5
ENTER VALUES OF X AND Y, ONE PAIR PER LINE
  2.5   18.75
  5.0    4.46
 10.0    1.502
 15.0    0.832
 20.0    0.566
```

(b) *on the graph-plotter*

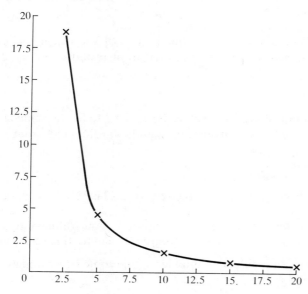

7.6 Graph of $y = f(x)$: J06EAF
[An easy routine]

The **purpose** of this routine is to draw a graph of $y = f(x)$ in a given specified interval $[a,b]$.

Specimen problem

To draw the graph

$$y = \frac{x + 3}{x(x - 2)}$$

for values of x in the interval $[-20,10]$.

The **routine name** with parameters is

J06EAF (F, X1, X2).

Description of parameters

Comment:
In common with most graphical routines, J06EAF has no output parameters. It also has no error parameter, which is very unusual.

Parameters which require values before J06EAF is called

X1⎫ [real variables]
Y1⎭ :

X1 and X2 should specify the interval [*a,b*] for which the graph is required. So, in the case of the specimen problem above, X1 = −20.0 and X2 = 10.0.

Comment:

Ref.
§7.2.2

Normally, you will want the values X1 and X2 to be the same as XMIN and XMAX. So you can arrange to read these values in to your program only once.

Functions which require definition

F: [a real function. It must be declared as EXTERNAL in your calling

Ref.
§2.2

program]

F is the function *f*, whose graph is required. You will have to supply a routine F(X) at the end of your program to define this function.

For the specimen problem, a suitable routine could be written as follows:

```
REAL FUNCTION F(X)
REAL X
F = (X + 3.0)/(X*(X - 2.0))
RETURN
END
```

Comment:

In the routine above, X is a parameter which gets its value from J06EAF (when necessary). All you are required to do here is to define the function F.

Specimen program

Program planning

1. *Declare*   REAL X1, X2, XMIN, XMAX, YMIN, YMAX
                INTEGER MARGIN
                EXTERNAL F

2. *Use* the framework in §7.2.4.

3. *Insert*
       Call   J06AAF (Axes)
       Set   X1, X2 (to XMIN and XMAX)
       Call   J06EAF
4. *Write routine* F(X)

Comments:

(i) The following program specifically draws the graph of

$$y = \frac{x + 3}{x(x - 2)}.$$

When you want to draw the graph of another function, you will have to change the function F accordingly.

(ii) It is possible that you won't know the maximum and minimum values of y for this graph. In this case, make a guess. You could then have a look at the resulting graph on the terminal using X04ABF. If the graph is satisfactory, then change X04ABF to PLOTS (or equivalent) and run the program again to obtain the graph from the graph-plotter.

(iii) You may need to make some changes to this program in order to make it run correctly on your computer. *See §3.4 for details.* You may also need to replace the statement

CALL PLOTS (0, 0, 5)

by one appropriate to your computer.

J06EAF specimen program

```
C       J06EAF: GRAPH OF FUNCTION
C       J06AAF: (WITH AXES)

        REAL X1, X2, XMIN, XMAX, YMIN, YMAX
        INTEGER MARGIN
        EXTERNAL F

        MARGIN = 1

        WRITE (6,*) 'ENTER THE LEAST AND GREATEST VALUES OF X'
        READ (5,*) XMIN, XMAX
        WRITE (6,*) 'ENTER THE LEAST AND GREATEST VALUES OF Y'
        READ (5,*) YMIN, YMAX

        CALL PLOTS(0,0,5)
        CALL J06WAF
        CALL J06WBF(XMIN,XMAX,YMIN,YMAX,MARGIN)
```

```
        CALL J06AAF

        X1 = XMIN
        X2 = XMAX

        CALL J06EAF(F,X1,X2)
```

```
        CALL J06WZF

        STOP

        END
```

continued

```
REAL FUNCTION F(X)
REAL X
F = (X+3.0)/(X*(X-2.0))
RETURN
END
```

J06EAF specimen run

(a) *on the terminal*

```
ENTER THE LEAST AND GREATEST VALUES OF X
-20.0  10.0
ENTER THE LEAST AND GREATEST VALUES OF Y
-10.0  10.0
```

(b) *on the graph-plotter*

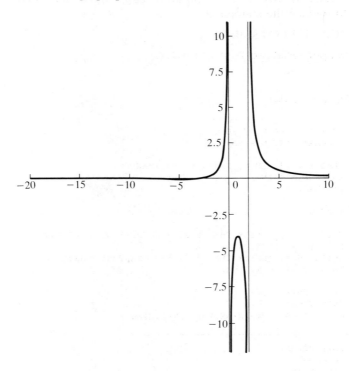

Postscript

Comment:

Note that the routine can deal with singularities. The specimen problem had singularities at $x = 2$ and $x = 0$. The routine arranges for asymptotes to be drawn at these points.

7.7 Contour map: J06GEF
[A medium routine]

The **purpose** of this routine is to draw a contour map of a given function over a specified region.

Specimen problem

To draw a contour map of the function

$$z(x,y) = e^{\frac{1}{2}x}(4x^2 + 4xy + 2y + 1)$$
over the region $-1 \leqslant x \leqslant 1, \quad -1 \leqslant y \leqslant 1$.

The **method** uses a contouring algorithm by K. J. Falconer.

The **routine name** with parameters is

J06GEF (HT, NX, NY, N, CHTS, ICH, IFAIL).

Description of parameters

Parameters which require values before J06GEF is called

N: [an integer variable]
N should contain the number of contours to be drawn.

For instance, in the specimen problem above, if contours were required for 9 different heights (i.e. 9 different z-values), then N should be given the value 9.

ICH: [an integer variable]
The actual heights of the contours can be chosen or determined in three different ways according to whether ICH is given the value 0, 1 or 2.

Value given to ICH	Way in which the contour heights are defined
ICH = 0	The routine automatically chooses the contour heights for you.
ICH = 1	You must supply the heights of the contours in the array CHTS.
ICH = 2	You supply two heights in CHTS(1) and CHTS(2). The routine then puts in N contours in this range.

CHTS: [A real one-dimensional array. Its length must be at least N in the
Ref. REAL declaration in your calling program]
p. 19 This array is used to specify the heights at which you want the contours drawn.

If you would like the routine to choose the N contour heights for you, then just put ICH = 0, and don't worry about CHTS before the routine is called.

If you would like to choose the contour heights yourself, then put ICH = 1, and the N chosen contour heights in CHTS(1), CHTS(2), . . ., CHTS(N).

If you are interested just in contours between two heights A and B, then put ICH = 2, CHTS(1) = A, and CHTS(2) = B. In this case, the routine itself will put in N equally spaced contours between these two heights (including A and B).

NX⎫ [both integer variables]
NY⎭:

NX and NY must specify the number of grid lines in the x and y directions respectively. The resulting mesh is used to find the contours.

Comments:
   (i) NX ≥ 2, NY ≥ 2 and NX*NY ≤ 150.
   (ii) The more mesh points you specify (subject to the restrictions above), the more accurate your contour map is likely to be.

So, in the specimen problem where $-1 \leq x \leq 1$ and $-1 \leq y \leq 1$, it seems reasonable to give the values NX = 12 and NY = 12.

Subroutines which require definition

HT: [a subroutine. It must be declared as EXTERNAL in your calling
Ref.   program]
§2.2

The **purpose** of this routine is to supply the function whose contour map is required.

The **name** of this routine with parameters is

   HT (X, Y, Z).

The way to use this subroutine is best demonstrated by example. If you wanted to write a subroutine to define the particular function in the specimen problem then a suitable subroutine HT could be written as follows:

```
SUBROUTINE HT (X, Y, Z)
REAL X, Y, Z
Z = EXP(0.5*X)*(4.0*X*X+4.0*X*Y+2.0*Y+1.0)
RETURN
END
```

Comment:

In the subroutine above, no values have been assigned to X or Y. This is because the NAG routine J06GEF supplies the required values of these variables. *In no circumstances should you attempt to assign values to X or Y in HT.*

The error parameter

IFAIL: [an integer variable]

IFAIL is the error parameter described in §3.2. It is recommended that you set

$$IFAIL = 0$$

before you call J06GEF. Then in the event of the routine failing, your program will stop and print one of the following error messages:

Error message	Meaning	Advice
IFAIL = 1	*Either* ICH ≠ 0, 1 or 2 *or* ICH = 2 and N < 2 *or* N < 1.	Check you have assigned sensible values to ICH and N.
IFAIL = 2	*Either* NX < 2 *or* NY < 2 *or* NX*NY > 150.	Check that your values of NX and NY conform with the restrictions.
IFAIL = 3	The routine is unable to find suitable contours.	Try increasing NX and NY. Choose a different routine. (See 'Associated routines'.)

Parameters to be examined after calling J06GEF

CHTS: [the real array described previously]

Before J06GEF is called, the array CHTS, used in conjunction with ICH, specifies the contour heights required. After a successful call of the routine, then whatever value was given to ICH initially, the array CHTS(1), CHTS(2), ..., CHTS(N) will contain the heights of the contours on the resulting map. *It is advisable to get the contour heights printed – especially in the event that the routine chooses your contour heights for you (i.e. ICH = 0). The specimen program arranges for the contour heights to be printed at the terminal.*

Specimen program

Program planning

1. *Declare*   REAL CHTS( ), XMIN, XMAX, YMIN, YMAX
   INTEGER NX, NY, N, ICH, IFAIL, MARGIN
   EXTERNAL HT

2. *Use* the framework in §7.2.5.

3. *Insert*

Call	J06AAF (Axes)
Set (or Read)	NX, NY, ICH
Read	N, (CHTS)
Set	IFAIL
Call	J06GEF
Print	CHTS (the contour heights)

4. *Write subroutine* HT.

Comments:

(i) The following program draws a contour map of the function specified in the specimen problem, with a printout of the contour heights at the terminal. In this case the contour heights are supplied by the routine. When you wish to draw a contour graph for a different function, you will have to change the subroutine HT to specify your particular function. If you require more than 20 contours, you will have to change the length of the array of CHTS in the REAL declaration in the program accordingly.

(ii) You may need to make some changes to this program in order to make it run correctly on your computer. *See §3.4 for details*. You may also need to replace the statement

      CALL PLOTS (0, 0, 5)

by one appropriate to your computer.

J06GEF specimen program

```
C       J06GEF: CONTOURS OF Z(X,Y)
C       J06AAF: (WITH AXES)

        REAL CHTS(20), XMIN, XMAX, YMIN, YMAX
        INTEGER NX, NY, N, ICH, IFAIL, MARGIN, I
        EXTERNAL HT

        MARGIN = 1

        WRITE (6,*) 'ENTER THE LEAST AND GREATEST VALUES OF X'
        READ (5,*) XMIN, XMAX
```

```
          WRITE (6,*) 'ENTER THE LEAST AND GREATEST VALUES OF Y'
          READ (5,*) YMIN, YMAX

          CALL PLOTS(0,0,5)
          CALL J06WAF
          CALL J06WBF(XMIN,XMAX,YMIN,YMAX,MARGIN)
```

```
          CALL J06AAF

          NX = 12
          NY = 12
          ICH = 0

          WRITE (6,*) 'ENTER NUMBER OF CONTOURS REQUIRED'
          READ (5,*) N

          IFAIL = 0

          CALL J06GEF(HT,NX,NY,N,CHTS,ICH,IFAIL)

          WRITE (6,*) 'CONTOUR NO.        HEIGHT'
          DO 10 I = 1, N
            WRITE (6,*) I, '                ', CHTS(I)
       10 CONTINUE
```

```
          CALL J06WZF

          STOP

          END

          SUBROUTINE HT(X,Y,Z)
          REAL X, Y, Z
          Z = EXP(0.5*X)*(4.0*X*X+4.0*X*Y+2.0*Y+1.0)
          RETURN
          END
```

J06GEF specimen run

(a) on the terminal

```
ENTER THE LEAST AND GREATEST VALUES OF X
-1.0    1.0
ENTER THE LEAST AND GREATEST VALUES OF Y
-1.0    1.0
ENTER NUMBER OF CONTOURS REQUIRED
9
CONTOUR NO.        HEIGHT
1                  -1.492400
2                  0.8168162
3                  3.126032
4                  5.435248
5                  7.744463
6                  10.05368
7                  12.36289
8                  14.67211
9                  16.98133
```

continued

(b) *on the graph-plotter*

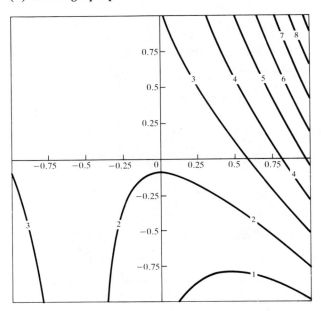

Postscript

Associated routines:

(i) If you want to draw a table of the contour heights alongside the contour map, then use J06GZF in conjunction with J06WCF.

(ii) If the routine described here is unable to find suitable contours, (IFAIL = 3), then try using J06GFF.

(iii) If you want a 3-D picture of your surface, use J06HCF.

The documentation of the routines above can be found in the *NAG Graphical Supplement*.

Part 3
The NAG Library

This part of the book consists of information which you need to know to enable you to use the official description of the routines in the *NAG Manual*. Chapter 10 consists of a series of descriptions which are taken from the *NAG Manual*, along with commentaries. These are arranged in order of difficulty, and can be used as practice exercises if the reader requires.

8
Approaching the NAG Library

First of all, you will have to ascertain where a copy of the *NAG Manual* can be found in your establishment.

The manual is a large affair, consisting of six volumes, and an additional volume describing the *Graphical Supplement*.

You will find the manual easier to use if you understand its structure. At the beginning of *Volume 1*, there is an Introduction to the Library, and a list of contents. The rest of the manual is divided into *Chapters*. Each chapter covers some particular area of numerical analysis or statistics, and consists of a *Chapter Introduction*, followed by a detailed description of each routine in that chapter.

As well as the full *NAG Manual*, most establishments hold copies of the *NAG Mini-Manual*. This single volume contains the necessary extracts from the full manual which enable you to choose the NAG routine best suited to your problem. Thus, if you have a numerical analysis or statistics problem, and want to know whether there is a suitable NAG routine to solve it, then an easy way to find out is to consult the *NAG Mini-Manual*. When you have chosen a suitable routine, then you would have to consult the full manual for its description, as the descriptions are not included in the *Mini-Manual*.

Another quick source of reference provided at some establishments is the *NAG On-Line Information Supplement*. This provides essential information at a terminal on which routine to use, and advice on how to use it. It also provides a description of the parameters for each routine and explanation of possible IFAIL values. The *NAG On-Line Information Supplement*, used in conjunction with the *Mini-Manual*, often provides sufficient information to successfully run a NAG routine. It is particularly useful if you have used the routine before. However, when you first use a NAG routine, you are recommended to read the description in the *NAG Manual*, as this contains the most detailed information generally available.

When you start to look for a suitable NAG routine to solve your problem, first you will have to find the appropriate chapter in the manual. The introduction to the chapters are always worth reading. As well as giving some mathematical background to the routines, they also give guidance to help you choose which routine to use. All this information can be found both in the *Mini-Manual* and in the full manual.

For instance, suppose you wanted to solve the linear simultaneous equations

$$3x_1 + x_2 + x_3 = 5$$
$$x_1 + 4x_2 + x_3 = 6$$
$$x_1 + x_2 + 5x_3 = 7.$$

The 'Contents', or the 'Keyword Index', refers you to Chapter F04. Amongst other things, the introduction to F04 includes guidance as to which of the current 26 listed routines is appropriate for your particular problem. As all 26 routines solve linear simultaneous equations, this guidance is invaluable. There are two *decision trees* to help you decide. One is labelled 'Black-box Routines' and the other, 'General Purpose Routines'. The general purpose routines are rather confusingly named. These are, in fact, special purpose routines, which should be used only when there is no suitable 'black-box' routine available to solve your problem.

The 'black-box' decision tree is reproduced below:

Decision Trees

If at any stage the answer to a question is 'Don't know' this should be read as 'No'.

The name of the routine (if any) that should be used to factorise the matrix A is given in brackets after the name of the routine for solving the equations.

(i) Black box Routines for Unique Solution of AX = B

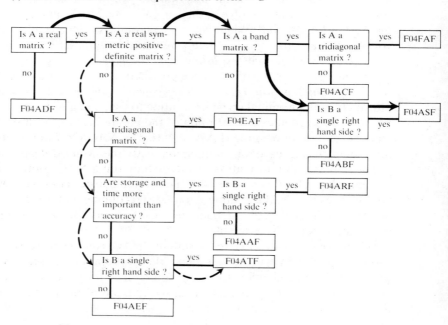

If you knew that the matrix

$$\begin{bmatrix} 3 & 1 & 1 \\ 1 & 4 & 1 \\ 1 & 1 & 5 \end{bmatrix}$$

was a symmetric positive-definite matrix, then you would follow the dark marked path on the decision tree, and come to the conclusion that you should use F04ASF. Had you not realized that the matrix was symmetric, then you would have followed the dotted path and landed up with F04ATF. This would, in fact, compute the solution to your problem. It just wouldn't do it as efficiently.

At this point, you are in a position to ask for the volume of the *NAG Library Manual* documentation which contains F04ASF. As you will need to study the document, and probably will not be allowed to take the manual away, make a copy of the document for your chosen routine.

The NAG documentation for F04ASF and further discussion about this routine can be found at the beginning of Chapter 9.

Summary

1. Find out where the *NAG Manual* and the *NAG Mini-Manual* are kept at your establishment.

2. Read the 'Introduction' at the beginning of the Manual or the Mini-Manual.

3. Find the chapter appropriate to your problem.

4. Read the introduction to this chapter.

5. Use the section 'Recommendations on choice and use of routines' to help you choose an appropriate routine.

6. Ask for the volume of the *NAG Library Manual* which contains the document for the routine of your choice.

7. Make a copy of this document for further study.

Comment:

Stages 1 and 2 above are really done only at the beginning of your NAG career. However, all the other stages are done *each* time you want to use a new NAG routine.

9
How to approach a NAG routine (F04ASF)

Most routines in the *NAG Manual* respond to roughly the same treatment. In this book, the NAG routine document F04ASF is used to demonstrate a method of approach.

9.1 F04ASF: Essential information

On the next three left-hand pages, you will find the documentation for F04ASF. It is suggested that you read this documentation first. You will discover that some of the information supplied there is essential, whereas some is less important. The right-hand pages are used for comments on the essential parts of the routine document.

F04ASF – NAG FORTRAN Library Routine Document

NOTE: before using this routine, please read the appropriate implementation document to check the interpretation of *bold italicised* terms and other implementation–dependent details. The routine name may be precision–dependent.

1.   Purpose   (9.1)

F04ASF calculates the accurate solution of a set of real symmetric positive definite linear equations with a single right hand side, $Ax = b$, by Cholesky's decomposition method.

2.   Specification

```
      SUBROUTINE F04ASF (A, IA, B, N, C, WK1, WK2, IFAIL)
C     INTEGER    IA, N, IFAIL
C     real       A(IA,N), B(N), C(N), WK1(N), WK2(N)
```
(9.2)

(9.3)

3.   Description

Given a set of linear equations $Ax = b$, where A is a real symmetric positive definite matrix, the routine uses Cholesky's method to decompose A into triangular matrices, $A = LL^T$ where L is lower triangular. An approximation to x is found by forward and backward substitution. The residual vector $r = b - Ax$ is then calculated and a correction, d, to x is found by the solution of $LL^T d = r$. x is then replaced by $(x+d)$ and the process repeated until machine accuracy is obtained. *Additional precision* accumulation of innerproducts is used throughout the calculation.

4.   References

[1]   WILKINSON, J.H. and REINSCH, C.
      Handbook for Automatic Computation.
      Volume II, Linear Algebra. pp. 31–34.
      Springer-Verlag, 1971.

5.   Parameters   (9.4)

A – *real* array of DIMENSION (IA,p) where $p \geq N$.

Before entry, A must contain the elements of the real symmetric positive definite matrix. The upper triangle only is needed.

On exit, the upper triangle will be unchanged, the strict lower triangle will be overwritten.

IA – INTEGER.

On entry, IA must specify the first dimension of array A as declared in the calling (sub)program.   (9.5)
$IA \geq N$.

Unchanged on exit.

B – *real* array of DIMENSION at least (N).

Before entry, B must contain the elements of the right-hand side. (See Section 11.) (9.6)
Unchanged on exit.   (9.7)

N – INTEGER.

On entry, N must specify the order of matrix A.

Unchanged on exit.

Commentary:

(9.1) *Purpose*: This confirms that the routine does the particular job which you require. In this case, the routine finds the unique solution of a set of linear simultaneous equations

$$Ax = b$$

where A is a positive-definite symmetric matrix.

Specification:
There is a wealth of information under this heading.

(9.2) The first piece of information supplied is the *name* of the routine (F04ASF) with a list of parameters needed to run this routine. This first statement in the specification also indicates the order in which the parameters must be written when the routine is called.

(9.3) Following the name of the routine, there is a statement (or statements) which indicate which parameters are real, which are integer, which are functions and so on.

Comment:
These 'type' statements in (9.3) are useful when it comes to constructing a program. This will be discussed further in §9.2.

(9.4) *Parameters*: It is essential to understand what each parameter does before trying to implement a particular routine. In an ideal world, you read through the description of the paramters, and understand it immediately . . . this seldom happens!

First, there is some fairly standard NAG jargon.

(9.5) *Before entry* ⎫
(9.6) *On entry*   ⎬

Effectively, any parameter whose description begins with these words has to be given a value before you call a NAG routine. Thus, in the case of F04ASF, A, IA, B, N, and IFAIL have to be given values before the routine is called.

C – *real* array of DIMENSION at least (N).

On successful exit, C will contain the solution vector.   (9.8)

WK1 – *real* array of DIMENSION at least (N).
WK2 – *real* array of DIMENSION at least (N).

Used as working space.

IFAIL – INTEGER.

Before entry, IFAIL must be set to 0 or 1. For users not familiar with this parameter (described in Chapter P01) the recommended value is 0.

Unless the routine detects an error (see next section), IFAIL contains 0 on exit.

6.   **Error Indicators and Warnings**   (9.9)
Errors detected by the routine:–

IFAIL = 1

The matrix A is not positive definite, possibly due to rounding errors.

IFAIL = 2

The refinement process fails to improve the solution, i.e. the matrix A is ill-conditioned.

7.   **Auxiliary Routines**
Details are distributed to sites in machine-readable form.

8.   **Timing**
The time taken is approximately proportional to N^3.

9.   **Storage**
There are no internally declared arrays.

10.   **Accuracy**   (9.10)
The computed solutions should be correct to full machine accuracy. For a detailed error analysis see [1], page 39.

11.   **Further Comments**
The routine **must not** be called with the same name for parameters B and C.

12.   **Keywords**
Accurate Solution of Linear Equations,
Cholesky Decomposition,
Real Symmetric Positive Definite Matrix,
Single Right Hand Side.

13.   **Example**   (9.11)
To solve the set of linear equations $Ax = b$ where

After a call to a routine, you need to know exactly what the routine has done to the value of each parameter.

(9.7) *Unchanged on exit*: This statement indicates that the value of this particular parameter is not changed by the routine.

(9.8) *On successful exit* (or sometimes '*on exit*'): This wording indicates either that the value of this parameter is changed by the routine, or that the routine has given this parameter a value.

For instance, in F04ASF, A has its value changed and C is given a value (namely, the solution).

It is very important that you take notice of these 'exit' messages, as sometimes routines destroy information which you might want to preserve. For instance, suppose that you needed to use the original contents of A after a call to F04ASF. Then you would have to make a copy of A before calling the routine, as this routine overwrites the contents of A.

(9.9) *Error indicators*: You should read these when the routine fails. If you don't understand the appropriate error message, then take your problem to a specialist.

(9.10) *Accuracy*: This section should always be read. It will usually give some indication as to how much accuracy there would be in your answer if the data were error-free. You should be warned, however, that important comments about the accuracy are sometimes found elsewhere in the routine description. So keep your eyes open for comments on accuracy.

$$A = \begin{pmatrix} 5 & 7 & 6 & 5 \\ 7 & 10 & 8 & 7 \\ 6 & 8 & 10 & 9 \\ 5 & 7 & 9 & 10 \end{pmatrix} \text{ and } b = \begin{pmatrix} 23 \\ 32 \\ 33 \\ 31 \end{pmatrix}$$

13.1. Program Text   (9.12)

WARNING: This **single precision** example program may require amendment for certain implementations. The results produced may not be the same. If in doubt, please seek further advice (see **Essential Introduction** to the Library Manual).

```
C       F04ASF EXAMPLE PROGRAM TEXT
C       NAG COPYRIGHT 1975
C       MARK 4.5 REVISED
C
        REAL A(6,6), B(5), C(5), WK1(18), WK2(4)
        INTEGER NIN, NOUT, I, N, J, IA, IFAIL
        DATA NIN /5/, NOUT /6/
        READ (NIN,99999) (WK1(I),I=1,7)
        WRITE (NOUT,99997) (WK1(I),I=1,6)
        N = 4
        READ (NIN,99998) ((A(I,J),J=1,N),I=1,N), (B(I),I=1,N)
        IA = 6
        IFAIL = 1
        CALL F04ASF(A, IA, B, N, C, WK1, WK2, IFAIL)
        IF (IFAIL.EQ.0) GO TO 20
        WRITE (NOUT,99996) IFAIL
        STOP
     20 WRITE (NOUT,99995) (C(I),I=1,N)
        STOP
99999 FORMAT (6A4, 1A3)
99998 FORMAT (4F5.0)
99997 FORMAT (4(1X/), 1H , 5A4, 1A3, 7HRESULTS/1X)
99996 FORMAT (25H0ERROR IN F04ASF IFAIL = , I2)
99995 FORMAT (10H0SOLUTIONS/(1H , F4.1))
        END
```

13.2. Program Data

```
F04ASF EXAMPLE PROGRAM DATA
    5    7    6    5
    7   10    8    7
    6    8   10    9
    5    7    9   10
   23   32   33   31
```

13.3. Program Results

```
F04ASF EXAMPLE PROGRAM RESULTS

SOLUTIONS
  1.0
  1.0
  1.0
  1.0
```

(9.11) *Example* ⎫
(9.12) *Example program* ⎭

These provide a way of checking that you really have understood the meaning of the parameters. Attached to each routine there is a program which can be used to solve the given example. Taking the example (9.11) in F04ASF, you can see that the order of this particular matrix is 4. Examining the example program, you will find the confirmation that the parameter N, which specifies the order of the matrix, is indeed set to 4. This kind of use of the example program can provide a great help in understanding the meaning of a parameter.

Comment:
You may be tempted to use the example programs given in the *NAG Manual* as a basis for a program to solve your own problem. That is a possible approach. However, one of the aims of this book is to encourage you to write your own programs in a different style. The specimen programs given in this book adopt a style which has several advantages (for most purposes) over the NAG example programs:

(a) They are written to deal with as general a problem as possible, consistent with clarity and simplicity.

(b) They are designed for use at a terminal, allowing you to enter your data interactively.

(c) They make use of some of the new features of Fortran 77. For example, they seldom bother with FORMAT statements. As a result they tend to be shorter than their NAG equivalents. Compare any program in this book with its sister program in the *NAG Manual* to see the difference. So, take courage in both hands, and write your own program.

Although the previous comments are attached specifically to the documentation of F04ASF, they are in fact quite general, and could be applied to almost any routine in the *NAG Manual*.

9.2 Housekeeping

There is a 'housekeeping' aspect to each routine. For instance, each time you tackle a new routine, you have to decide exactly what to do with each parameter. You have to decide

   (i) which parameters need to be declared,
  (ii) which parameters require values before the routine is called,
 (iii) which parameters should be printed after the routine is called,
 (iv) whether subroutines or functions have to be written

and so on.

Comments:

   (i) There is a great advantage in declaring all the parameters, independent of whether the declaration is essential or not. In this way, you minimize the chance of omitting to declare a parameter which must be declared. This declaration is *nearly* done for you in (9.3) in the routine description. However, do not be misled by the statement

     REAL A(IA,N), B(N), C(N), WK1(N), WK2(N).

Fortran requires that, in a main program, arrays have fixed sizes. So IA and N have to be replaced by numbers in your declaration. In fact, the N and IA in the statement above effectively indicate the *minimum* size required for these arrays. So if you wrote a declaration

     REAL A(4,4), B(4), C(4), WK1(4), WK2(4)

in your program, then you would be able to solve the specimen problem (9.11). However, it is good practice to declare more space than this, so that you can use the same program to solve larger sets of equations. So, make a decision as to the maximum number of equations you are normally likely to solve. If you decide, for instance, that normally you solve less than 20 equations in 20 unknowns, then suitable declarations would be

     INTEGER IA, N, IFAIL
     REAL A(20,20), B(20), C(20), WK1(20), WK2(20).

   (ii) As discussed in §9.1, the parameters requiring values before the routine is called are N, A, B, IA, and IFAIL. However, the way in which N, A, and B are dealt with is different from IA and IFAIL. Each time you want to use this program, the number of equations you want to solve, the left-hand side coefficients and the right-hand sides may be different. So this information should be entered each time before the routine is called.

However, IFAIL and IA do not require to be entered each time. If you decide to give IFAIL the value zero before you call the routine on one occasion, it seems very likely that you will want to do the same thing the next time that you run the program.

So, you would have a statement

IFAIL = 0

in your program, rather than read in the value of IFAIL each time.

Also, once you have made the declaration

REAL A(20,20), B(20), . . . ,

then you have fixed the value of IA. With this particular declaration, IA must be 20, so once again, you would have a statement

IA = 20

in your program.

(iii) In the event of a successful call of F04ASF, initially only C (the answer) will be of interest. If IFAIL is initially set to zero, the routine itself will print an error message if it fails.

(iv) This particular routine doesn't need any functions or subroutines.

To *summarize*, the following table contains most of the information which is needed to construct a program which calls F04ASF. The suggestions given in the right-hand column are based on the assumption that there are not more than 20 equations to be solved. If you have more equations than this, then you would have to change these suggestions, the subsequent program plan and the specimen program accordingly.

	Parameter	Meaning	Value needed before a routine call	To be printed after a routine call	Suggestions: size of array in declaration	Initial value
	N	Number of equations	✓			
	A	Left-hand side coefficients	✓		A(20,20)	
Ref. §2.4	IA	First dimension of A in declaration	✓			IA=20
	B	Right-hand sides of equations	✓		B(20)	
	C	The solution **x** in A**x** = **b**		✓	C(20)	
Ref. §3.3	WK1 WK2	Arrays used as workspace			WK1(20) WK2(20)	
Ref. §3.2	IFAIL	The error parameter	✓			IFAIL=0

Comment:
When planning how to use a routine, it is not necessary to go through the formality of making a table like the one above. However, it is necessary to have extracted this information in some form or another.

After this, the process of constructing a basic program is a reasonably simple matter.

9.3 Specimen program

If the 'housekeeping' job has been thoroughly done, there are few problems left in planning and writing a program. Putting together the information from the table above, the program can be planned as follows:

Program plan

1. *Declare* INTEGER IA, N, IFAIL
   REAL A(20,20), B(20), C(20), WK1(20), WK2(20)

2. *Read* N, A, B

3. *Set* IA=20, IFAIL=0

4. *Call* F04ASF

5. *Print* C

At this point, all that remains is to fill in the details of the program.

Comment:
The exact order of some of the statements in the following program is largely a matter of aesthetic choice.

F04ASF specimen program

```
C       F04ASF: LINEAR SIMULTANEOUS EQUATIONS
C               (SYMMETRIC POSITIVE-DEFINITE MATRIX)

        REAL A(20,20), B(20), C(20), WK1(20), WK2(20)
        INTEGER IA, N, IFAIL, I, J

        IA = 20

        WRITE (6,*) 'ENTER THE NUMBER OF EQUATIONS'
        READ (5,*) N
        WRITE (6,*) 'ENTER THE MATRIX A'
        DO 10 I = 1, N
          READ (5,*) (A(I,J),J=1,N)
10      CONTINUE
        WRITE (6,*)
     *    'ENTER THE RIGHT-HAND SIDE COEFFICIENTS,
     *    'ONE PER LINE'
```

```
      DO 20 I = 1, N
         READ (5,*) B(I)
   20 CONTINUE

      IFAIL = 0

      CALL F04ASF(A,IA,B,N,C,WK1,WK2,IFAIL)

      WRITE (6,*) 'THE SOLUTION IS'
      DO 30 I = 1, N
         WRITE (6,*) C(I)
   30 CONTINUE

      STOP

      END
```

F04ASF specimen run

```
ENTER THE NUMBER OF EQUATIONS
 4
ENTER THE MATRIX A
 5.0    7.0    6.0    5.0
 7.0   10.0    8.0    7.0
 6.0    8.0   10.0    9.0
 5.0    7.0    9.0   10.0
ENTER THE RIGHT-HAND SIDE COEFFICIENTS, ONE PER LINE
 23.0
 32.0
 33.0
 31.0
THE SOLUTION IS
1.000000
1.000000
1. 00000
1.000000
```

9.4 Chapter summary

Each time you want to implement a NAG routine, you will need to do the following:

(a) Understand the description of each of the parameters.
(b) Decide what is the maximum size problem you want to run.
(c) Do the 'Housekeeping' exercise described in §9.2.
(d) Write a program plan.
(e) Write and adequately test your program.

10
Some specimen NAG routine documents

10.1 Introduction

In this final chapter, the planning ideas of the previous chapter are applied to some other routines. It is hoped that this will demonstrate to you an overall strategy for coping with virtually all the routines in the NAG Library. The routines in this chapter have been chosen for different reasons. Some cover areas not discussed elsewhere in the text, others cover programming points not presented previously.

This time, the routines are arranged roughly in order of the difficulty of their use, starting with the easiest!

Each of the following descriptions are in three parts: Part I consists of a description of the routine taken from the *NAG Manual*. When you have read this, take the opportunity to do your own housekeeping, and to write your own specimen programs. You will then be able to compare these with the programs which you will find in Part III. Part II contains explanatory comments on specific parts of the description which might have caused difficulty.

At this point, dear reader, I wish you the best of luck in your adventures with the NAG Library. I am sure that you will find it a worthwhile and rewarding experience.

Note:

The NAG routine descriptions, reproduced in Part I of each of the following sections, have been re-typeset for this book, and may look superficially different from the same descriptions in your copy of the *NAG Manual*. However, the essential information has been reproduced exactly.

10.2 Eigenvalues and eigenvectors (symmetric matrix)

(I)

F02ABF – NAG FORTRAN Library Routine Document

NOTE: before using this routine, please read the appropriate implementation document to check the interpretation of *bold italicised* terms and other implementation–dependent details. The routine name may be precision–dependent.

1. **Purpose**

    F02ABF calculates all the eigenvalues and eigenvectors of a real symmetric matrix by Householder reduction and the QL algorithm.

2. **Specification**

   ```
         SUBROUTINE F02ABF (A, IA, N, R, V, IV, E, IFAIL)
   C     INTEGER    IA, N, IV, IFAIL
   C     real       A(IA,N), R(N), V(IV,N), E(N)
   ```

3. **Description**

    This routine reduces the real symmetric matrix A to a real symmetric tridiagonal matrix by Householder's method. The eigenvalues and eigenvectors are calculated using the QL algorithm.

4. **References**

    [1] WILKINSON, J.H. and REINSCH, C.
         Handbook for Automatic Computation.
         Volume II, Linear Algebra, pp. 212-226 and 227-240.
         Springer-Verlag, 1971.

5. **Parameters**

    A – *real* array of DIMENSION (IA,p) where p $\geq$ N.
    Before entry, A must contain the elements of the symmetric matrix; the lower triangle only is needed.

    Unchanged on exit, but see Section 11.

    IA – INTEGER.
    On entry, IA must specify the first dimension of array A as declared in the calling (sub)program.
    IA $\geq$ N.
    Unchanged on exit.

    N – INTEGER.
    On entry, N must specify the order of the matrix.
    Unchanged on exit.

    R – *real* array of DIMENSION at least (N).
    On successful exit, R contains the eigenvalues in ascending order.

    V – *real* array of DIMENSION (IV,q) where q $\geq$ N.
    On successful exit, V contains the normalised eigenvectors in columns corresponding to the eigenvalues i.e. V(J,I), J = 1,2,...,N corresponds to R(I).

    IV – INTEGER.
    On entry, IV must specify the first dimension of array V as declared in the calling (sub)program.

$IV \geq N$.

Unchanged on exit.

E – *real* array of DIMENSION at least (N).

Used as working space.

IFAIL – INTEGER.

Before entry, IFAIL must be set to 0 or 1. For users not familiar with this parameter (described in Chapter P01) the recommended value is 0.

Unless the routine detects an error (see next section), IFAIL contains 0 on exit.

6.   Error Indicators and Warnings

Errors detected by the routine:–

IFAIL = 1

Failure in F02AMF indicating that more than $30 \times N$ iterations are required to isolate all the eigenvalues.

7.   Auxiliary Routines

Details are distributed to sites in machine-readable form.

8.   Timing

The time taken is approximately proportional to N^3.

9.   Storage

There are no internally declared arrays.

10.   Accuracy

The eigenvectors are always accurately orthogonal but the accuracy of the individual eigenvectors is dependent on their inherent sensitivity to changes in the original matrix. For a detailed error analysis see [1], pages 222 and 235.

11.   Further Comments

The eigenvectors are normalised so that the sum of squares of the elements is equal to 1. If the subroutine is called with the same name for the arrays A and V the eigenvectors will overwrite the original matrix.

12.   Keywords

Eigenvalues and Eigenvectors,
Householder Reduction,
QL Algorithm,
Real Symmetric Matrix.

13.   Example

To calculate all the eigenvalues and eigenvectors of the real symmetric matrix:

$$\begin{pmatrix} 5 & 4 & 1 & 1 \\ 4 & 5 & 1 & 1 \\ 1 & 1 & 4 & 2 \\ 1 & 1 & 2 & 4 \end{pmatrix}$$

13.1. Program Text

WARNING: This **single precision** example program may require amendment for certain implementations. The results produced may not be the same. If in doubt, please seek further advice (see **Essential Introduction** to the Library Manual).

```
C      FØ2ABF EXAMPLE PROGRAM TEXT
C      NAG COPYRIGHT 1975
C      MARK 4.5 REVISED
C
       REAL A(5,5), R(5), V(5,5), E(7)
       INTEGER NIN, NOUT, I, N, J, IA, IV, IFAIL
       DATA NIN /5/, NOUT /6/
       READ (NIN,99999) (E(I),I=1,7)
       WRITE (NOUT,99997) (E(I),I=1,6)
       N = 4
       READ (NIN,99998) ((A(I,J),J=1,N),I=1,N)
       IA = 5
       IV = 5
       IFAIL = 1
       CALL FØ2ABF(A, IA, N, R, V, IV, E, IFAIL)
       IF (IFAIL.EQ.Ø) GO TO 20
       WRITE (NOUT,99996) IFAIL
       STOP
    20 WRITE (NOUT,99995) (R(I),I=1,N)
       WRITE (NOUT,99994) ((V(I,J),J=1,N),I=1,N)
       STOP
99999 FORMAT (6A4, 1A3)
99998 FORMAT (4F5.Ø)
99997 FORMAT (4(1X/), 1H , 5A4, 1A3, 7HRESULTS/1X)
99996 FORMAT (25HØERROR IN FØ2ABF IFAIL = , I2)
99995 FORMAT (12HØEIGENVALUES/(1H , 11X, 4F9.1))
99994 FORMAT (13HØEIGENVECTORS/(1H , 12X, 4F9.4))
       END
```

13.2. Program Data

```
FØ2ABF EXAMPLE PROGRAM DATA
   5    4    1    1
   4    5    1    1
   1    1    4    2
   1    1    2    4
```

13.3. Program Results

```
FØ2ABF EXAMPLE PROGRAM RESULTS

EIGENVALUES
                1.Ø      2.Ø      5.Ø     1Ø.Ø

EIGENVECTORS
              Ø.7Ø71   Ø.ØØØØ  -Ø.3162   Ø.6325
             -Ø.7Ø71   Ø.ØØØØ  -Ø.3162   Ø.6325
              Ø.ØØØØ   Ø.7Ø71   Ø.6325   Ø.3162
              Ø.ØØØØ  -Ø.7Ø71   Ø.6325   Ø.3162
```

(II) Commentary

Summary of information needed to run F02ABF

Purpose: This routine finds all the eigenvalues and a set of corresponding eigenvectors of a real symmetric matrix.

> *Comment*:
> The routine takes advantage of the *symmetry* of the matrix A in two ways:
> (a) it economizes on computing time and storage, since, roughly speaking, it only needs to work with half the matrix.
> (b) the eigenvalues and eigenvectors are known to be real (as opposed to complex).
>
> The result is that this routine is easier to use than the more general routine F02AGF, described in §5.10.

The **routine name** with parameters is

F02ABF (A, IA, N, R, V, IV, E, IFAIL).

Housekeeping

	Parameter	Meaning	Value needed before a routine call	To be printed after a routine call	Suggestions: size of array in declaration	Initial value
	N [Integer]	The order of A	✓			
	A [Real array]	The matrix whose eigenvalues are required	✓		A(10,10)	
Ref. §2.4	IA [Integer]	First dimension of A in declaration	✓			IA=10
	R [Real array]	The eigenvalues		✓	R(10)	
	V [Real array]	Corresponding eigenvectors		✓	V(10,10)	
	IV [Integer]	First dimension of V in declaration	✓			IV=10
Ref. §3.3	E [Real array]	Array used as workspace			E(10)	
Ref. §3.2	IFAIL [Integer]	The usual error parameter	✓			IFAIL=0

> *Comments*:
> (i) The suggestions given in the right-hand column of the table above

are based on the assumption that A (the matrix whose eigenvalues and eigenvectors are required) has a maximum order 10. For larger matrices, this column, the subsequent program plan and the specimen program would have to be changed accordingly.

(ii) Eigenvectors corresponding to each of the eigenvalues are stored in the columns of V. So, for example, an eigenvector corresponding to the third eigenvalue R(3) will be found in V(1,3), V(2,3), . . ., V(N,3).

(III) Program

Program plan

1. *Declare* REAL A(10,10), R(10), V(10,10), E(10)
   INTEGER IA, N, IV, IFAIL

2. *Read*   N, A

3. *Set*   IA=10; IV=10; IFAIL=0

4. *Call*   F02ABF

5. *Print* R, V

The following program is used to find the eigenvalues and corresponding eigenvectors of the matrix given in the NAG routine document.

i.e.
$$A = \begin{bmatrix} 5 & 4 & 1 & 1 \\ 4 & 5 & 1 & 1 \\ 1 & 1 & 4 & 2 \\ 1 & 1 & 2 & 4 \end{bmatrix}$$

F02ABF specimen program

Ref.
§5.2

In the following program the data are read from a data file.

```
C       F02ABF: EIGENVALUES AND EIGENVECTORS
C               (SYMMETRIC MATRIX)

        REAL A(10,10), R(10), V(10,10), E(10)
        INTEGER IA, N, IV, IFAIL, I, J

        OPEN (UNIT=20,FILE='F02ABF.DAT')

        IA = 10
        IV = 10

        WRITE (6,*) 'THE ORDER OF THE MATRIX'
        READ (20,*) N
        WRITE (6,*) N
        WRITE (6,*) 'THE MATRIX A'
        DO 10 I = 1, N
           READ (20,*) (A(I,J),J=1,N)
           WRITE (6,'(10F5.2)') (A(I,J),J=1,N)
     10 CONTINUE

        IFAIL = 0
```

continued

```
CALL F02ABF(A,IA,N,R,V,IV,E,IFAIL)

DO 30 I = 1, N
   WRITE (6,*)
   WRITE (6,*) 'AN EIGENVALUE IS'
   WRITE (6,'(E15.4)') R(I)
   WRITE (6,*) 'WITH CORRESPONDING EIGENVECTOR'
   DO 20 J = 1, N
      WRITE (6,'(E15.4)') V(J,I)
20    CONTINUE
30 CONTINUE

STOP

END
```

F02ABF specimen run

```
THE ORDER OF THE MATRIX
4
THE MATRIX A
5.00  4.00  1.00  1.00
4.00  5.00  1.00  1.00
1.00  1.00  4.00  2.00
1.00  1.00  2.00  4.00

AN EIGENVALUE IS
     0.1000E+01
WITH CORRESPONDING EIGENVECTOR
     0.7071E+00
    -0.7071E+00
     0.1990E-07
    -0.1147E-07

AN EIGENVALUE IS
     0.2000E+01
WITH CORRESPONDING EIGENVECTOR
    -0.1724E-07
     0.1675E-07
     0.7071E+00
    -0.7071E+00

AN EIGENVALUE IS
     0.5000E+01
WITH CORRESPONDING EIGENVECTOR
    -0.3162E+00
    -0.3162E+00
     0.6325E+00
     0.6325E+00

AN EIGENVALUE IS
     0.1000E+02
WITH CORRESPONDING EIGENVECTOR
     0.6325E+00
     0.6325E+00
     0.3162E+00
     0.3162E+00
```

10.3 Linear simultaneous equations (complex coefficients)

(I) <center>**F04ADF – NAG FORTRAN Library Routine Document**</center>

NOTE: before using this routine, please read the appropriate implementation document to check the interpretation of **bold italicised** terms and other implementation–dependent details. The routine name may be precision–dependent.

1. **Purpose**

    F04ADF calculates the approximate solution of a set of complex linear equations with multiple right hand sides by Crout's factorisation method.

2. **Specification**

    ```
    SUBROUTINE FØ4ADF (A, IA, B, IB, N, M, C, IC, WKSPCE, IFAIL)
    C    complex    A(IA,N), B(IB,M), C(IC,M)
    C    INTEGER    IA, IB, N, M, IC, IFAIL
    C    real       WKSPCE(N)
    ```

3. **Description**

    Given a set of complex linear equations $AX = B$, the routine first decomposes A using Crout's factorisation with partial pivoting, $PA = LU$, where P is a permutation matrix, L is lower triangular and U is unit upper triangular. The columns x of the solution X are found by forward and backward substitution in $Ly = Pb$ and $Ux = y$ where b is a column of the right hand side matrix B. *Additional precision* accumulation of innerproducts is used throughout.

4. **References**

    [1]   WILKINSON, J.H. and REINSCH, C.
           Handbook for Automatic Computation.
           Volume II, Linear Algebra, pp. 93-110.
           Springer-Verlag, 1971.

5. **Parameters**

    A – *complex* array of DIMENSION (IA,p) where $p \geq N$.

    Before entry, A must contain the elements of the complex matrix.

    On successful exit, it contains the Crout factorisation with the unit diagonal of U understood.

    IA – INTEGER.

    On entry, IA must specify the first dimension of array A as declared in the calling (sub)program.
    $IA \geq N$
    Unchanged on exit.

    B – *complex* array of DIMENSION (IB,q) where $q \geq M$.

    Before entry, B must contain the elements of the M right hand sides stored in columns.

    Unchanged on exit, but see Section 11.

    IB – INTEGER.

    On entry, IB must specify the first dimension of array B as declared in the calling (sub)program.
    $IB \geq N$
    Unchanged on exit.

N – INTEGER.
>   On entry, N must specify the order of matrix A.
>
>   Unchanged on exit.

M – INTEGER.
>   On entry, M must specify the number of right hand sides.
>
>   Unchanged on exit.

C – *complex* array of DIMENSION (IC,r) where r $\geq$ M.
>   On successful exit, C contains the M complex solution vectors.

IC – INTEGER.
>   On entry, IC must specify the first dimension of array C as declared in the calling (sub)program.
>   IC $\geq$ N
>
>   Unchanged on exit.

WKSPCE – *real* array of DIMENSION at least (N).
>   Used as working space.

IFAIL – INTEGER.
>   Before entry, IFAIL must be set to 0 or 1. For users not familiar with this parameter (described in Chapter P01) the recommended value is 0.
>
>   Unless the routine detects an error (see next section), IFAIL contains 0 on exit.

6.   Error Indicators and Warnings

Errors detected by the routine:–

IFAIL = 1
>   Failure in F03AHF, the matrix A is singular, possibly due to rounding errors.

7.   Auxiliary Routines

Details are distributed to sites in machine-readable form.

8.   Timing

The time taken is approximately proportional to N^3.

9.   Storage

There are no internally declared arrays.

10.   Accuracy

The accuracy of the computed solution depends on the conditioning of the original matrix. For a detailed error analysis see [1], page 106.

11.   Further Comments

If the routine is called with the same name for parameters B and C, then the solution vectors will overwrite the right hand sides.

12.   Keywords

Approximate Solution of Linear Equations,
Complex Matrix,
Crout Factorisation,
Multiple Right Hand Sides.

13.   Example

To solve the set of linear equations $AX = B$ where

$$A = \begin{pmatrix} 1 & 1+2i & 2+10i \\ 1+i & 3i & -5+14i \\ 1+i & 5i & -8+20i \end{pmatrix} \quad \text{and} \quad B = \begin{pmatrix} 1 \\ 0 \\ 0 \end{pmatrix}$$

13.1.   Program Text

WARNING: This **single precision** example program may require amendment for certain implementations. The results produced may not be the same. If in doubt, please seek further advice (see **Essential Introduction** to the Library Manual).

```
C      F04ADF EXAMPLE PROGRAM TEXT
C      NAG COPYRIGHT 1975
C      MARK 4.5 REVISED
C
       COMPLEX A(10,10), B(10,1), C(10,1)
       REAL WKSPCE(18)
       INTEGER NIN, NOUT, I, N, M, J, IA, IB, IC, IFAIL
       DATA NIN /5/, NOUT /6/
       READ (NIN,99999) (WKSPCE(I),I=1,7)
       WRITE (NOUT,99997) (WKSPCE(I),I=1,6)
       N = 3
       M = 1
       READ (NIN,99998) ((A(I,J),J=1,N),I=1,N), (B(I,1),I=1,N)
       IA = 10
       IB = 10
       IC = 10
       IFAIL = 1
       CALL F04ADF(A, IA, B, IB, N, M, C, IC, WKSPCE, IFAIL)
       IF (IFAIL.EQ.0) GO TO 20
       WRITE (NOUT,99996) IFAIL
       STOP
    20 WRITE (NOUT,99995) (C(I,1),I=1,N)
       STOP
99999 FORMAT (6A4, 1A3)
99998 FORMAT (6(F4.0, 2X))
99997 FORMAT (4(1X/), 1H , 5A4, 1A3, 7HRESULTS/1X)
99996 FORMAT (25H0ERROR IN F04ADF IFAIL = , I2)
99995 FORMAT (10H0SOLUTIONS/(2H (, F5.1, 1H,, F5.1, 1H)))
       END
```

13.2.   Program Data

```
F04ADF EXAMPLE PROGRAM DATA
  1.0    0.0    1.0    2.0    2.0   10.0
  1.0    1.0    0.0    3.0   -5.0   14.0
  1.0    1.0    0.0    5.0   -8.0   20.0
  1.0    0.0    0.0    0.0    0.0    0.0
```

F04ADF   *F04 – Simultaneous Linear Equations*

13.3. Program Results

```
F04ADF EXAMPLE PROGRAM RESULTS

SOLUTIONS
( 10.0,   1.0)
(  9.0,  -3.0)
( -2.0,   2.0)
```

(II) Commentary

Summary of information needed to run F04ADF

Purpose: This routine finds the unique solution of sets of linear simultaneous equations, where the left-hand side coefficients and the right-hand sides are complex. There can be more than one right-hand side.

Comment:
This routine is included to illustrate the use of COMPLEX arrays to store complex data.

The **routine name** with parameters is

F04ADF (A, IA, B, IB, N, M, C, IC, WKSPCE, IFAIL).

Housekeeping

	Parameter	Meaning	Value needed before a routine call	To be printed after a routine call	Suggestions: size of array in declaration	Initial value
	N [Integer]	The number of equations	✓			
	M [Integer]	The number of right-hand sides	✓			
	A [Complex array]	The left-hand side coefficients of AC = B	✓		A(10,10)	
Ref. §2.4	IA [Integer]	First dimension of A in declaration	✓			IA=10
	B [Complex array]	The right-hand sides B of AC = B	✓		B(10,5)	
	IB [Integer]	First dimension of B in declaration	✓			IB=10
	C [Complex array]	The solution C in AC = B		✓	C(10,5)	
	IC [Integer]	First dimension of C in declaration	✓			IC=10
Ref. §3.3	WKSPCE [Real array]	Array used as workspace			WKSPCE(10)	
Ref. §3.2	IFAIL [Integer]	The usual error parameter	✓			IFAIL=0

Comments:

(i) The suggestions given in the right-hand column of the table above would allow for a maximum of 10 equations with up to 5 right-hand sides. If you have more equations, or more than 5 right-hand sides, then you would have to change this column, the subsequent program plan and the specimen program accordingly.

Ref.
§2.3
(ii) As each element of the arrays A, B, and C have to contain a number pair (x,y) corresponding to the complex number $x + iy$, then these arrays must be declared as COMPLEX.

(iii) The solution corresponding to the i(th) right-hand side will be found in the i(th) column of C. So, for example, if there is only one right-hand side, the solution will be found in $C(1,1), C(2,1), \ldots, C(N,1)$.

(III) Program

Program plan

1. *Declare* COMPLEX A(10,10), B(10,5), C(10,5)
   REAL WKSPCE(10)
   INTEGER IA, IB, N, M, IC, IFAIL

2. *Read*   N, M, A, B

3. *Set*   IA=10; IB=10; IC=10; IFAIL=0

4. *Call*   F04ADF

5. *Print*   C

The following program is used to solve the equations given in the NAG routine document.

i.e.
$$x_1 + (1 + 2i)x_2 + (2 \quad + 10i)x_3 = 1$$
$$(1 + i)x_1 + 3i \qquad x_2 + (-5 + 14i)x_3 = 0$$
$$(1 + i)x_1 + 5i \qquad x_2 + (-8 + 20i)x_3 = 0.$$

F04ADF specimen program

Ref.
§5.2
In the following program the data are read from a data file.

```
C       F04ADF: LINEAR SIMULTANEOUS EQUATIONS
C               (COMPLEX COEFFICIENTS)

        COMPLEX A(10,10), B(10,5), C(10,5)
        REAL WKSPCE(10)
        INTEGER IA, IB, N, M, IC, IFAIL, I, J

        OPEN (UNIT=20,FILE='F04ADF.DAT')

        IA = 10
        IB = 10
        IC = 10
```

```
      WRITE (6,*) 'THE NUMBER OF EQUATIONS'
      READ (20,*) N
      WRITE (6,*) N
      WRITE (6,*) 'THE NUMBER OF RIGHT HAND SIDES'
      READ (20,*) M
      WRITE (6,*) M
      WRITE (6,*) 'THE MATRIX A'
      DO 10 I = 1, N
        READ (20,*) (A(I,J),J=1,N)
        WRITE (6,90) (A(I,J),J=1,N)
   10 CONTINUE
      WRITE (6,*) 'THE MATRIX B'
      DO 20 I = 1, N
        READ (20,*) (B(I,J),J=1,M)
        WRITE (6,90) (B(I,J),J=1,M)
   20 CONTINUE

      IFAIL = 0

      CALL F04ADF(A,IA,B,IB,N,M,C,IC,WKSPCE,IFAIL)

      WRITE (6,*) 'THE SOLUTIONS ARE'
      DO 30 I = 1, N
        WRITE (6,91) I, (C(I,J),J=1,M)
   30 CONTINUE

      STOP

   90 FORMAT (1X,8(F5.2,' + ',F5.2,' I',3X))
   91 FORMAT (1X,'C(',I2,') = ',5(F5.2,' + ',F5.2,' I',3X))

      END
```

F04ADF specimen run

```
THE NUMBER OF EQUATIONS
3
THE NUMBER OF RIGHT HAND SIDES
1
THE MATRIX A
 1.00 +  0.00 I    1.00 +  2.00 I    2.00 + 10.00 I
 1.00 +  1.00 I    0.00 +  3.00 I   -5.00 + 14.00 I
 1.00 +  1.00 I    0.00 +  5.00 I   -8.00 + 20.00 I
THE MATRIX B
 1.00 +  0.00 I
 0.00 +  0.00 I
 0.00 +  0.00 I
THE SOLUTIONS ARE
C( 1) = 10.00 +  1.00 I
C( 2) =  9.00 + -3.00 I
C( 3) = -2.00 +  2.00 I
```

Comments:

(i) In the specimen program above, the left-hand-side coefficients and the right-hand sides of the equations are read from a data file. The data in the data file should be arranged in the following way:

```
3
1
(1.0,0.0)    (1.0,2.0)     (2.0,10.0)
(1.0,1.0)    (0.0,3.0)    (-5.0,14.0)
(1.0,1.0)    (0.0,5.0)    (-8.0,20.0)
(1.0,0.0)
(0.0,0.0)
(0.0,0.0)
```

(ii) Note that when using (∗) format in READ statements in Fortran 77, then a complex number $x + iy$ *must be* entered as a number pair (x,y), enclosed in brackets, as demonstrated in the data above.

10.4 Solution of non-linear simultaneous equations

(I)   **C05NBF – NAG FORTRAN Library Routine Document**

NOTE: before using this routine, please read the appropriate implementation document to check the interpretation of *bold italicised* terms and other implementation–dependent details. The routine name may be precision–dependent.

1. Purpose

C05NBF is an easy-to-use routine to find a zero of a system of N nonlinear functions in N variables by a modification of the Powell hybrid method.

2. Specification

```
    SUBROUTINE C05NBF (FCN, N, X, FVEC, XTOL, WA, LWA, IFAIL)
C       INTEGER N, LWA, IFAIL
C       real X(N), FVEC(N), XTOL, WA(LWA)
C       EXTERNAL FCN
```

3. Description

C05NBF is based upon the MINPACK routine HYBRD1. It chooses the correction at each step as a convex combination of the Newton and scaled gradient directions. Under reasonable conditions this guarantees global convergence for starting points far from the solution and a fast rate of convergence. The Jacobian is updated by the rank–1 method of Broyden. At the starting point the Jacobian is approximated by forward differences, but these are not used again until the rank–1 method fails to produce satisfactory progress.

4. References

[1] POWELL, M.J.D.
      A hybrid method for nonlinear algebraic equations. In 'Numerical Methods for Nonlinear Algebraic Equations', Ed. Rabinowitz, P., Gordon and Breach, 1970.

[2] MORE, J.J., GARBOW, B.S. and HILLSTROM, K.E.
      User Guide for MINPACK-1.
      ANL–80–74, Argonne National Laboratory.

5. Parameters

FCN – SUBROUTINE, supplied by the user.

FCN must calculate the values of the functions at X and return these in the vector FVEC.
Its specification is:–

```
    SUBROUTINE FCN(N, X, FVEC, IFLAG)
    INTEGER N, IFLAG
    real X(N), FVEC(N)
```

N – INTEGER.

On entry, N contains the number of equations.
The value of N must not be changed by FCN.

X – *real* array of DIMENSION (N).

On entry, X contains the point at which the functions are to be evaluated.
The values in X must not be changed by FCN.

FVEC – *real* array of DIMENSION (N).

On exit, unless IFLAG is reset to a negative number, FVEC(i) must contain the value of the (i)th function evaluated at X.

IFLAG – INTEGER.

In general, IFLAG should not be reset by FCN. If, however, the user wishes to terminate execution (perhaps because some illegal point X has been reached) then IFLAG should be set to a negative integer. This value will be returned through IFAIL.

FCN must be declared as EXTERNAL in the calling (sub)program.

N – INTEGER.

On entry, N must specify the number of equations.
N > 0.

Unchanged on exit.

X – *real* array of DIMENSION at least (N).

Before entry, X(j) must be set to a guess at the j(th) component of the solution (j = 1,2,...,N).

On exit, X contains the final estimate of the solution vector.

FVEC – *real* array of DIMENSION at least (N).

On exit, FVEC contains the function values at the final point, X.

XTOL – *real*.

On entry, XTOL must specify the accuracy in X to which the solution is required. XTOL ≥ 0.0. The recommended value is the square root of machine precision.

Unchanged on exit.

WA – *real* array of DIMENSION at least (LWA).

Used as workspace.

LWA – INTEGER.

On entry, LWA must specify the dimension of the array WA.
LWA $\geq \frac{1}{2}$N×(3×N+13).

Unchanged on exit.

IFAIL – INTEGER.

On entry, IFAIL must be set to 0 or 1. For users not familiar with this parameter (described in Chapter P01) the recommended value is 0.

Unless the routine detects an error (see next section), IFAIL contains 0 on exit.

6.   Error Indicators and Warnings

Errors detected by the routine:–

IFAIL < 0

This indicates an exit from C05NBF because the user has set IFLAG negative in FCN. The value of IFAIL will be the same as the user's setting of IFLAG.

IFAIL = 1

> On entry, $N \leq 0$,
> or $\quad XTOL < 0.0$,
> or $\quad LWA < \frac{1}{2}N \times (3 \times N + 13)$.

IFAIL = 2

> There have been at least $200 \times (N + 1)$ evaluations of FCN. Consider restarting the calculation from the final point held in X.

IFAIL = 3

> No further improvement in the approximate solution X is possible; XTOL is too small.

IFAIL = 4

> The iteration is not making good progress. This failure exit may indicate that the system does not have a zero, or that the solution is very close to the origin (see Section 10). Otherwise, rerunning C05NBF from a different starting point may avoid the region of difficulty. Alternatively consider using C05PBF or C05PCF which require an analytic Jacobian.

7. Auxiliary Routines

Details are distributed to sites in machine-readable form.

8. Timing

The time required by C05NBF to solve a given problem depends on N, the behaviour of the functions, the accuracy requested and the starting point. The number of arithmetic operations executed by C05NBF to process each call of FCN is about $11.5 \times N^2$. Unless FCN can be evaluated quickly, the timing of C05NBF will be strongly influenced by the time spent in FCN.

9. Storage

There are no internally declared arrays.

10. Accuracy

C05NBF tries to ensure that

$$||X - XSOL||_2 \leq XTOL \times ||XSOL||_2.$$

If this condition is satisfied with $XTOL = 10^{-k}$ then the larger components of X have k significant decimal digits. There is a danger that the smaller components of X may have large relative errors, but the fast rate of convergence of C05NBF usually avoids this possibility.

If XTOL is less than machine precision (see NAG Library routine X02AAF), and the above test is satisfied with the machine precision in place of XTOL, then the routine exits with IFAIL = 3.

Note that this convergence test is based purely on relative error, and may not indicate convergence if the solution is very close to the origin.

The test assumes that the functions are reasonably well behaved. If this condition is not satisfied, then C05NBF may incorrectly indicate convergence. The validity of the answer can be checked, for example, by rerunning C05NBF with a tighter tolerance.

11. Further Comments

Ideally the problem should be scaled so that at the solution the function values are of comparable magnitude.

12.   Keywords

Equations, nonlinear algebraic, easy-to-use;
Powell Hybrid Method, easy-to-use.

13.   Example

To determine the values $x_1, ..., x_9$ which satisfy the tridiagonal equations:–

$$(3-2x_1)x_1 - 2x_2 = -1$$

$$-x_{i-1} + (3-2x_i)x_i - 2x_{i+1} = -1, \quad i = 2,3,...,8$$

$$-x_8 + (3-2x_9)x_9 = -1$$

13.1.   Program Text

WARNING: This **single precision** example program may require amendment for certain implementations. The results produced may not be the same. If in doubt, please seek further advice (see **Essential Introduction** to the Library Manual).

```
C       C05NBF EXAMPLE PROGRAM TEXT
C       MARK 9 RELEASE. NAG COPYRIGHT 1981
C       .. LOCAL SCALARS ..
        REAL FNORM, TOL
        INTEGER IFAIL, J, NOUT
C       .. LOCAL ARRAYS ..
        REAL FVEC(9), WA(180), X(9)
C       .. FUNCTION REFERENCES ..
        REAL F05ABF, SQRT, X02AAF
C       .. SUBROUTINE REFERENCES ..
C       C05NBF
C       ..
        EXTERNAL FCN
        DATA NOUT /6/
        WRITE (NOUT,99999)
C       THE FOLLOWING STARTING VALUES PROVIDE A ROUGH SOLUTION.
        DO 20 J=1,9
           X(J) = -1.E0
     20 CONTINUE
        TOL = SQRT(X02AAF(0.0))
        IFAIL = 0
        CALL C05NBF(FCN, 9, X, FVEC, TOL, WA, 180, IFAIL)
        FNORM = F05ABF(FVEC,9)
        WRITE (NOUT,99998) FNORM, IFAIL, (X(J),J=1,9)
        STOP
99999 FORMAT (4(1X/), 31H C05NBF EXAMPLE PROGRAM RESULTS/1X)
99998 FORMAT (5X, 31H FINAL L2 NORM OF THE RESIDUALS, E12.4//5X,
     *  15H EXIT PARAMETER, I10//5X, 27H FINAL APPROXIMATE SOLUTION//
     *  (5X, 3E12.4))
        END
        SUBROUTINE FCN(N, X, FVEC, IFLAG)
C       .. SCALAR ARGUMENTS ..
        INTEGER IFLAG, N
C       .. ARRAY ARGUMENTS ..
        REAL FVEC(N), X(N)
C       ..
C       .. LOCAL SCALARS ..
        REAL ONE, TEMP, TEMP1, TEMP2, THREE, TWO, ZERO
        INTEGER K
C       ..
        DATA ZERO, ONE, TWO, THREE /0.E0,1.E0,2.E0,3.E0/
```

```
      DO 20 K=1,N
         TEMP = (THREE-TWO*X(K))*X(K)
         TEMP1 = ZERO
         IF (K.NE.1) TEMP1 = X(K-1)
         TEMP2 = ZERO
         IF (K.NE.N) TEMP2 = X(K+1)
         FVEC(K) = TEMP - TEMP1 - TWO*TEMP2 + ONE
   20 CONTINUE
      RETURN
      END
```

13.2.   Program Data

None.

13.3.   Program Results

```
    C05NBF EXAMPLE PROGRAM RESULTS

       FINAL L2 NORM OF THE RESIDUALS   0.1193E-07

       EXIT PARAMETER          0

       FINAL APPROXIMATE SOLUTION

       -0.5707E+00  -0.6816E+00  -0.7017E+00
       -0.7042E+00  -0.7014E+00  -0.6919E+00
       -0.6658E+00  -0.5960E+00  -0.4164E+00
```

(II) Commentary

Summary of information needed to run C05NBF

Purpose: This routine attempts to find a solution of a set of non-linear simultaneous equations, or in other words, a zero of a set of non-linear functions.

Comment:

The problem is to find values $x_1, x_2, \ldots, x_n$ such that

$$f_1(x_1, x_2, \ldots, x_n) = 0$$
$$f_2(x_1, x_2, \ldots, x_n) = 0$$
.
.
.
$$f_n(x_1, x_2, \ldots, x_n) = 0$$

where $f_1, f_2, \ldots, f_n$ are known non-linear functions.

The **routine name** with parameters is

C05NBF (FCN, N, X, FVEC, XTOL, WA, LWA, IFAIL).

Housekeeping

	Parameter	Meaning	Value needed before a routine call	To be printed after a routine call	Suggestions: size of array in declaration	Initial value
	N [Integer]	The number of equations	✓			
Ref. §2.2	FCN [EXTERNAL subroutine]	A subroutine defining the equations	See comment (ii) below			
	X [Real array]	(a) Initial guess at a solution	✓		X(10)	
		(b) A solution		✓		
	FVEC [Real array]	The function values at the solution		✓	FVEC(10)	
	XTOL [Real]	The accuracy required in the solution	✓			XTOL= SQRT(X02AAF(0.0))
Ref. §3.3	WA [Real array]	Array used workspace			WA(215)	
Ref. §2.4	LWA [Integer]	The length of WA in the declaration	✓			LWA=215
Ref. §3.2	IFAIL [Integer]	The usual error parameter	✓			IFAIL=0

Comments:

(i) The suggestions given in the right-hand column of the table above would allow for a maximum of 10 equations in 10 unknowns. If you have more equations, then you would have to change this column, the subsequent program plan, and the specimen program accordingly.

(ii) The subroutine FCN is the place where you describe the equations you want to solve.

Suppose, for instance, that you want to solve the equations

$$x_1^2 + 2x_2^2 - 35.58 = 0$$
$$x_1 + x_2 + \cos(x_1 x_2) - 5.279 = 0,$$

then a suitable subroutine FCN could be written as follows:

```
SUBROUTINE FCN (N, X, FVEC, IFLAG)
INTEGER N, IFLAG
REAL X(N), FVEC(N)
FVEC(1) = X(1)**2 + 2.0*X(2)**2 - 35.58
FVEC(2) = X(1) + X(2) + COS (X(1)*X(2)) - 5.279
RETURN
END
```

There are some important points to note about this subroutine:

(a) FCN must be declared as EXTERNAL in the declaration in the calling program.

(b) As N is a parameter of the subroutine FCN, you can use the declaration

REAL X(N), FVEC(N)

in FCN.

(c) The only thing which you do in this subroutine is to set up the equations to be solved.

Take heed of the warning in the NAG routine document not to assign values to N or X in the subroutine FCN. The routine C05NBF supplies these values when required.

(iii) The NAG routine document indirectly recommends you to set

XTOL = SQRT (X02AAF(0.0)).

This will produce the best possible accuracy in your answer.

(iv) There is no guarantee that a set of non-linear equations has a solution, or that this routine will always find a solution even if there is one from an arbitrary initial guess. If you get IFAIL = 2 or IFAIL = 4, check the coding of your subroutine FCN and try different initial guesses for X.

(III) Program

Program plan

1. *Declare*	REAL X(10), FVEC(10), XTOL, WA(215) INTEGER N, LWA, IFAIL EXTERNAL FCN
2. *Read*	N, X
3. *Set*	LWA=215; XTOL=SQRT (X02AAF(0.0)); IFAIL=0
4. *Call*	C05NBF
5. *Print*	X, FVEC

6. *Write a subroutine* FCN

The following program is used to solve the equations given in the NAG routine document:

i.e. $(3 - 2x_1)x_1 - 2x_2 = -1$
$-x_{i-1} + (3 - 2x_i)x_i - 2x_{i+1} = -1$ $i = 2, 3, \ldots, 8$
$-x_8 + (3 - 2x_9)x_9 = -1$

C05NBF specimen program

```
C       C05NBF: NON-LINEAR SIMULTANEOUS EQUATIONS

        REAL X(10), FVEC(10), XTOL, WA(215), X02AAF
        INTEGER N, LWA, IFAIL, I
        EXTERNAL FCN

        XTOL = SQRT(X02AAF(0.0))
        LWA = 215

        WRITE (6,*) 'ENTER THE NUMBER OF EQUATIONS'
        READ (5,*) N
        WRITE (6,*) 'ENTER A GUESS AT THE SOLUTION'
        READ (5,*) (X(I),I=1,N)

        IFAIL = 0

        CALL C05NBF(FCN,N,X,FVEC,XTOL,WA,LWA,IFAIL)

        WRITE (6,*)
     *   'AN ESTIMATED SOLUTION TO THE EQUATIONS IS'
        DO 10 I = 1, N
           WRITE (6,*) 'X(', I, ') = ', X(I)
 10     CONTINUE
        WRITE (6,*) 'WITH CORRESPONDING FUNCTION VALUES'
        DO 20 I = 1, N
           WRITE (6,*) 'F(', I, ') = ', FVEC(I)
 20     CONTINUE
```

```
          STOP

          END

          SUBROUTINE FCN(N,X,FVEC,IFLAG)
          INTEGER N, IFLAG, I
          REAL X(N), FVEC(N)
          FVEC(1) = (3.0-2.0*X(1))*X(1) - 2.0*X(2) + 1.0
          DO 10 I = 2, 8
             FVEC(I) = -X(I-1) + (3.0-2.0*X(I))*X(I) - 2.0*X(I+1)
        *                + 1.0
       10 CONTINUE
          FVEC(9) = -X(8) + (3.0-2.0*X(9))*X(9) + 1.0
          RETURN
          END
```

C05NBF specimen run

```
ENTER THE NUMBER OF EQUATIONS
 9
ENTER A GUESS AT THE SOLUTION
 -1.0  -1.0  -1.0  -1.0  -1.0  -1.0  -1.0  -1.0  -1.0
AN ESTIMATED SOLUTION TO THE EQUATIONS IS
X(1) = -0.5706534
X(2) = -0.6816280
X(3) = -0.7017328
X(4) = -0.7042137
X(5) = -0.7013699
X(6) = -0.6918661
X(7) = -0.6657919
X(8) = -0.5960338
X(9) = -0.4164123
WITH CORRESPONDING FUNCTION VALUES
F(1) = 5.1259995E-06
F(2) = 1.7881393E-06
F(3) = -9.2387199E-07
F(4) = -2.5033951E-06
F(5) = -2.9802322E-06
F(6) = -2.0712614E-06
F(7) = 2.9802322E-07
F(8) = 2.6747584E-06
F(9) = -1.6987324E-06
```

10.5 Multiple integral (with constant limits)

(I)

D01GBF – NAG FORTRAN Library Routine Document

NOTE: before using this routine, please read the appropriate implementation document to check the interpretation of *bold italicised* terms and other implementation–dependent details. The routine name may be precision–dependent.

1. Purpose

D01GBF returns an approximation to the integral of a function over a hyper-rectangular region, using a Monte-Carlo method. An approximate relative error estimate is also returned. This routine is suitable for low accuracy work.

2. Specification

```
      SUBROUTINE D01GBF (NDIM, A, B, MINCLS, MAXCLS,
     1                   FUNCTN, EPS, ACC, LENWRK, WRKSTR, FINEST, IFAIL)
C     INTEGER          NDIM, MINCLS, MAXCLS, LENWRK, IFAIL
C     real             A(NDIM), B(NDIM), EPS, ACC, WRKSTR(LENWRK),
C     1                FINEST
C     EXTERNAL         FUNCTN
```

3. Description

D01GBF uses an adaptive Monte-Carlo method based on the algorithm described by Lautrup [1]. It is implemented for integrals of the form:

$$\int_{a_1}^{b_1} \int_{a_2}^{b_2} \dots \int_{a_n}^{b_n} F(x) \, dx_n \dots dx_2 dx_1$$

Upon entry, unless LENWRK has been set to the minimum value $10 \times$ NDIM, the routine subdivides the integration region into a number of equal volume subregions. Inside each subregion the integral and the variance are estimated by means of pseudo-random sampling. All contributions are added together to produce an estimate for the whole integral and total variance. The variance along each co-ordinate axis is determined and the routine uses this information to increase the density and change the widths of the subintervals along each axis, so as to reduce the total variance. The total number of subregions is then increased by a factor of two and the program recycles for another iteration. The program stops when a desired accuracy has been reached or too many integral evaluations are needed for the next cycle.

4. References

[1]   LAUTRUP, B.
       An Adaptive Multi-dimensional Integration Procedure.
       Proc. 2nd Coll. on Advanced Methods in Theoretical Physics.
       Marseille, 1971.

5. Parameters

NDIM – INTEGER.

On entry, NDIM must specify the number of dimensions of the integral.
NDIM ≥ 1.

Unchanged on exit.

A – *real* array of DIMENSION at least (NDIM).

On entry, A(I) must be set to the Ith lower limit of the multiple integral, for I = 1,...,NDIM.

Unchanged on exit.

B – *real* array of DIMENSION at least (NDIM).

> On entry, B(I) must be set to the Ith upper limit of the multiple integral, for I = 1,...,NDIM.
>
> Unchanged on exit.

MINCLS – INTEGER.

> Before entry, MINCLS must be set:
>
> either to the minimum number of integrand evaluations to be allowed, in which case MINCLS ≥ 0;
>
> or to a negative value. In this case the routine assumes that a previous call had been made with the same parameters NDIM, A and B and with either the same integrand (in which case D01GBF continues that calculation) or a similar integrand (in which case D01GBF begins the calculation with the subdivision used in the last iteration of the previous call). See also WRKSTR.
>
> On exit, MINCLS contains the number of integrand evaluations actually used by D01GBF.

MAXCLS – INTEGER.

> On entry, MAXCLS must be set to the maximum number of integrand evaluations to be allowed. In the continuation case this is the number of new integrand evaluations to be allowed. These counts do not include zero integrand values.
> MAXCLS ≥ 4 × (NDIM + 1).
>
> Unchanged on exit.

FUNCTN – *real* FUNCTION, supplied by the user.

> FUNCTN must evaluate the integrand at a specified point.
>
> Its specification is:

```
real FUNCTION      FUNCTN(NDIM,X)
INTEGER            NDIM
real               X(NDIM)
```

NDIM – INTEGER.

> On entry, NDIM specifies the number of dimensions of the integral, as given in the call of D01GBF. NDIM must not be reset.

X – *real* array of DIMENSION (NDIM).

> On entry, X contains the NDIM co-ordinates of the point at which the integrand is to be evaluated. X must not be reset.
>
> FUNCTN must be declared as EXTERNAL in the calling (sub)program.

EPS – *real*.

> On entry, EPS must be set to the relative accuracy required.
> EPS ≥ 0.0.
>
> Unchanged on exit.

ACC – *real*.

> On exit, ACC specifies the estimated relative accuracy of FINEST.

LENWRK – INTEGER.

On entry, LENWRK must be set to the length of the array WRKSTR as declared in the calling (sub)program.
LENWRK $\geq 10 \times$ NDIM.
For maximum efficiency, LENWRK should be about

$$3 \times \text{NDIM} \times (\text{MAXCLS}/4)^{1/\text{NDIM}} + 7 \times \text{NDIM}.$$

If LENWRK is given the value $10 \times$ NDIM then the subroutine uses only one iteration of a crude Monte-arlo method with MAXCLS sample points.
Unchanged on exit.

WRKSTR – *real* array of DIMENSION (LENWRK).

Used as working storage. If on entry MINCLS < 0.0, then WRKSTR must be unchanged from the previous call of D01GBF – except that for a new integrand WRKSTR(LENWRK) must be set to 0.0. See MINCLS.

On exit, WRKSTR contains information about the current subinterval structure which could be used in later calls of D01GBF. In particular, WRKSTR(J) gives the number of subintervals used along the Jth co-ordinate axis.

FINEST – *real*.

On exit, FINEST contains the best estimate obtained for the integral.

IFAIL – INTEGER.

Before entry, IFAIL must be set to 0 or 1. For users not familiar with this parameter (described in Chapter P01) the recommended value is 0.

Unless the routine detects an error (see next section), IFAIL contains 0 on exit.

6.   Error Indicators and Warnings

Errors detected by the routine:

IFAIL = 1

On entry, NDIM < 1
  or MINCLS $\geq$ MAXCLS
  or LENWRK $< 10 \times$ NDIM
  or MAXCLS $< 4 \times (\text{NDIM} + 1)$
  or EPS < 0.0

IFAIL = 2

MAXCLS was too small for D01GBF to obtain the required relative accuracy EPS. In this case D01GBF returns a value of FINEST with estimated relative error ACC, but ACC will be greater than EPS. This error exit may be taken before MAXCLS non-zero integrand evaluations have actually occurred, if the routine calculates that the current estimates could not be improved before MAXCLS was exceeded.

7.   Auxiliary Routines

Details are distributed to sites in machine-readable form.

8.   Timing

The running time for D01GBF will usually be dominated by the time used to evaluate the integrand FUNCTN, so the maximum time that could be used is approximately proportional to MAXCLS.

9.   Storage

There are no internally declared arrays.

10.   Accuracy

A relative error estimate is output through the parameter ACC. The confidence factor is set so that the actual error should be less than ACC 90% of the time. If a user desires a higher confidence level then a smaller value of EPS should be used.

11.   Further Comments

For some integrands, particularly those that are poorly behaved in a small part of the integration region, D01GBF may terminate with a value of ACC which is significantly smaller than the actual relative error. This should be suspected if the returned value of MINCLS is small relative to the expected difficulty of the integral. Where this occurs, D01GBF should be called again, but with a higher entry value of MINCLS (e.g. twice the returned value) and the results compared with those from the previous call.

The exact values of FINEST and ACC on return will depend (within statistical limits) on the sequence of random numbers generated within D01GBF by calls to the NAG routine G05CAF. Separate runs will produce identical answers unless the part of the program executed prior to calling D01GBF also calls (directly or indirectly) routines from the G05 chapter, and the series of such calls differs between runs. If desired, the user may ensure the identity or difference between runs of the results returned by D01GBF, by calling the NAG routine G05CBF or G05CCF (qq.v.) respectively, immediately before calling D01GBF.

12.   Keywords

Adaptive Integration,
Monte-Carlo Method,
Multiple Integral.

13.   Example

This example program calculates the integral

$$\int_0^1 \int_0^1 \int_0^1 \int_0^1 \frac{4x_1 x_3^2 \exp(2x_1 x_3)}{(1+x_2+x_4)^2} dx_1 dx_2 dx_3 dx_4 = 0.575364$$

13.1.   Program Text

WARNING: This **single precision** example program may require amendment for certain implementations. The results produced may not be the same. If in doubt, please seek further advice (see **Essential Introduction** to the Library Manual).

```
C       D01GBF EXAMPLE PROGRAM TEXT
C       MARK 10 RELEASE. NAG COPYRIGHT 1982.
C       .. LOCAL SCALARS ..
        REAL ACC, EPS, FINEST
        INTEGER IFAIL, K, MAXCLS, MINCLS, NDIM, NOUT
C       .. LOCAL ARRAYS ..
        REAL A(4), B(4), WRKSTR(500)
C       .. SUBROUTINE REFERENCES ..
C       D01GBF
C       ..
        EXTERNAL FUNCTN
        DATA NOUT /6/
        WRITE (NOUT,99999)
        NDIM = 4
        DO 20 K=1,NDIM
```

```
            A(K) = 0.0
            B(K) = 1.0
   20 CONTINUE
      EPS = .01
      MAXCLS = 20000
      MINCLS = 1000
      IFAIL = 1
      CALL D01GBF(NDIM, A, B, MINCLS, MAXCLS, FUNCTN, EPS, ACC,
     * 500, WRKSTR, FINEST, IFAIL)
      IF (IFAIL.EQ.0) GO TO 40
      WRITE (NOUT,99998) IFAIL
      GO TO 60
   40 WRITE (NOUT,99997) EPS, FINEST, ACC, MINCLS
   60 STOP
99999 FORMAT (4(1X/), 31H D01GBF EXAMPLE PROGRAM RESULTS/1X)
99998 FORMAT (22H0D01GBF FAILS. IFAIL =, I2)
99997 FORMAT (5X, 23HREQUESTED ACCURACY   = , E13.5/5X, 9HESTIMATED,
     * 14H VALUE        = , E13.5/5X, 23HESTIMATED ACCURACY   = ,
     * E13.5/5X, 23HNUMBER OF EVALUATIONS =, I11)
      END
      REAL FUNCTION FUNCTN(NDIM, X)
C     .. SCALAR ARGUMENTS ..
      INTEGER NDIM
C     .. ARRAY ARGUMENTS ..
      REAL X(NDIM)
C     ..
C     .. FUNCTION REFERENCES ..
      REAL EXP
C     ..
      FUNCTN = 4.0*X(1)*X(3)**2*EXP(2.0*X(1)*X(3))/(1.0+X(2)+X(4))**
     * 2
      RETURN
      END
```

13.2. Program Data
None.

13.3. Program Results

```
D01GBF EXAMPLE PROGRAM RESULTS

    REQUESTED ACCURACY   =   0.10000E-01
    ESTIMATED VALUE      =   0.57554E+00
    ESTIMATED ACCURACY   =   0.81989E-02
    NUMBER OF EVALUATIONS =      1728
```

(II) Commentary

Summary of information needed to run D01GBF

Purpose: The routine finds an estimate for an *n*-dimensional integral.

$$\int_{a_1}^{b_1}\int_{a_2}^{b_2} \ldots \ldots \int_{a_n}^{b_n} f(\mathbf{x}) \mathrm{d}x_n \ldots \mathrm{d}x_2 \mathrm{d}x_1$$

where the lower and upper limits of the integral are constants.

The **routine name** with parameters is

D01GBF (NDIM,A,B,MINCLS,MAXCLS,FUNCTN,EPS,ACC,
LENWRK,WRKSTR,FINEST,IFAIL).

Housekeeping

	Parameter	Meaning	Value needed before a routine call	To be printed after a routine call	Suggestions: size of array in declaration	Initial value
	NDIM [Integer]	The number of dimensions in the integral	✓			
	A,B [Real arrays]	Arrays holding the lower and upper limits	✓		A(10) B(10)	
	MINCLS [Integer]	Minimum number of integrand evaluations	✓			MINCLS =1000
	MAXCLS [Integer]	Maximum number of integrand evaluations	✓			MAXCLS =20,000
Ref. **§2.2**	FUNCTN [EXTERNAL function routine]	A function routine in which the function to be integrated is defined	See comment (iii) below			
Ref. **§5.4** **(b)**	EPS [Real]	The relative accuracy required in the answer	✓			
	ACC [Real]	The estimated relative accuracy in the answer		✓		

continued

Table—continued

	Parameter	Meaning	Value needed before a routine call	To be printed after a routine call	Suggestions: size of array in declaration	Initial value
Ref. §3.3	WRKSTR [Real array]	Array used as workspace			WRKSTR(500)	
Ref. §2.4	LENWRK [Integer]	Length of array WRKSTR in declaration	✓			LENWRK=500
	FINEST [Real]	The answer		✓		
Ref. §3.2	IFAIL [Integer]	The usual error parameter	✓			IFAIL=0

Comments:

(i) The suggestions in the right-hand column of the table above would allow for integrals in up to 10 dimensions. If your integral has more dimensions than this, then you would have to change this column, the subsequent program plan and the specimen program accordingly.

(ii) If you are not getting the accuracy you requested (IFAIL=2), then try increasing MAXCLS.

(iii) A function routine (FUNCTN), specifying the particular function to be integrated

i.e. $\dfrac{4x_1x_3^2 \exp(2x_1x_3)}{(1 + x_2 + x_4)^2}$,

could be written as follows:

```
REAL FUNCTION FUNCTN(NDIM,X)
INTEGER NDIM
REAL X(NDIM)
FUNCTN=4.0*X(1)*X(3)**2*EXP(2.0*X(1)*X(3))/(1.0+X(2)
*          +X(4))**2
RETURN
END
```

There are two important points to note about this function routine:

(a) FUNCTN must be declared as EXTERNAL in the calling program.

(b) As NDIM is a parameter of FUNCTN, you can use the declaration

REAL X(NDIM)

in FUNCTN.

(iv) A minimum declaration WRKSTR(100) is needed in a program which evaluates integrals in up to 10 dimensions. However, the suggestion is that you are likely to get better results if you use rather more workspace. Initially, you could take the advice of the specimen program, and use REAL WRKSTR(500).

(III) Program

Program plan

1. *Declare*	REAL A(10), B(10), EPS, ACC, WRKSTR(500), FINEST INTEGER NDIM, MINCLS, MAXCLS, LENWRK, IFAIL EXTERNAL FUNCTN
2. *Read*	NDIM, A, B, EPS
3. *Set*	LENWRK=500; MINCLS=1000; MAXCLS=20000; IFAIL=0
4. *Call*	D01GBF
5. *Print*	FINEST, ACC

6. *Write a function routine* FUNCTN.

The following program is used to evaluate the integral given in the NAG routine document:

i.e.   Calculate the integral

$$\int_0^1 \int_0^1 \int_0^1 \int_0^1 \frac{4x_1 x_3^2 \exp(2x_1 x_3)}{(1 + x_2 + x_4)^2} \, dx_4 dx_3 dx_2 dx_1 \; .$$

D01GBF specimen program

```
C       D01GBF: MULTIPLE INTEGRAL
C               (CONSTANT LIMITS)

        REAL A(10), B(10), EPS, ACC, WRKSTR(500), FINEST
        INTEGER NDIM, MINCLS, MAXCLS, LENWRK, IFAIL, I
        EXTERNAL FUNCTN

        MINCLS = 1000
        MAXCLS = 20000
        LENWRK = 500

        WRITE (6,*)
     *   'ENTER THE NUMBER OF DIMENSIONS OF THE INTEGRAL'
        READ (5,*) NDIM
        WRITE (6,*) 'ENTER THE LOWER AND UPPER LIMITS'
```

continued

```
            WRITE (6,*) '(FROM THE OUTSIDE, ONE PAIR PER LINE)'
            DO 10 I = 1, NDIM
              READ (5,*) A(I), B(I)
      10 CONTINUE
            WRITE (6,*) 'ENTER THE RELATIVE ACCURACY'
            READ (5,*) EPS

            IFAIL = 0

            CALL D01GBF(NDIM,A,B,MINCLS,MAXCLS,FUNCTN,EPS,ACC,
           *           LENWRK,WRKSTR,FINEST,IFAIL)

            WRITE (6,*) 'ESTIMATED VALUE IS ', FINEST
            WRITE (6,*) 'ESTIMATED RELATIVE ACCURACY IS ', ACC

            STOP

            END

            REAL FUNCTION FUNCTN(NDIM,X)
            INTEGER NDIM
            REAL X(NDIM)
            FUNCTN = 4.0*X(1)*(X(3))**2*EXP(2.0*X(1)*X(3))
           *        /(1.0+X(2)+X(4))**2
            RETURN
            END
```

D01GBF specimen run

```
      ENTER THE NUMBER OF DIMENSIONS OF THE INTEGRAL
      4
      ENTER THE LOWER AND UPPER LIMITS
      (FROM THE OUTSIDE, ONE PAIR PER LINE)
      0.0  1.0
      0.0  1.0
      0.0  1.0
      0.0  1.0
      ENTER THE RELATIVE ACCURACY
      1.0E-3
      ESTIMATED VALUE IS 0.5751200
      ESTIMATED RELATIVE ACCURACY IS 8.3250032E-04
```

10.6 Linear simultaneous equations (symmetric positive-definite band matrix)

(I)

F04ACF – NAG FORTRAN Library Routine Document

NOTE: before using this routine, please read the appropriate implementation document to check the interpretation of *bold italicised* terms and other implementation–dependent details. The routine name may be precision–dependent.

1. Purpose

F04ACF calculates the approximate solution of a set of real symmetric positive definite band equations with multiple right hand sides by Cholesky's decomposition method.

2. Specification

```
       SUBROUTINE F04ACF (A, IA, B, IB, N, M, IR, C, IC, RL, IRL, M1,
      1           IFAIL)
C      INTEGER    IA, IB, N, M, IR, IC, IRL, M1, IFAIL
C      real       A(IA,M1), B(IB,IR), C(IC,IR), RL(IRL,M1)
```

3. Description

Given a set of linear equations $AX = B$, where A is a real symmetric positive definite band matrix, the routine uses Cholesky's method to decompose A into triangles such that $A = LL^T$, where L is a lower triangular band matrix. The columns x of the solution X are found by forward and backward substitution in $Ly = b$ and $L^T x = y$ where b is a column of the right hand side matrix B.

4. References

[1]   WILKINSON, J.H. and REINSCH, C.
      Handbook for Automatic Computation.
      Volume II, Linear Algebra, pp. 50-56.
      Springer-Verlag, 1971.

5. Parameters

A – *real* array of DIMENSION (IA,p) where $p \geq M1$.

Before entry, A must contain the elements of the lower half of the positive definite symmetric band matrix of order N, with M lines on either side of the diagonal, $A(I,M1)$, $I = 1,2,...,N$ being the diagonal elements of the matrix. The relationship between the stored array and the conventional array in the case $N=5$, $M=2$ is illustrated below.

Stored array

x	x	a_{13}
x	a_{22}	a_{23}
a_{31}	a_{32}	a_{33}
a_{41}	a_{42}	a_{43}
a_{51}	a_{52}	a_{53}

Lower Triangle of conventional array

a_{13}
a_{22} a_{23}
a_{31} a_{32} a_{33}
$\quad\quad a_{41}$ a_{42} a_{43}
$\quad\quad\quad\quad a_{51}$ a_{52} a_{53}

The elements in the upper left-hand corner of the stored array may be arbitrary since these are not used.

Unchanged on exit.

IA – INTEGER.

On entry, IA must specify the first dimension of array A as declared in the calling (sub)program.
$IA \geq N$
Unchanged on exit.

B – *real* array of DIMENSION (IB,p) where $p \geq IR$.

Before entry, B must contain the elements of the IR right hand sides stored in columns.
Unchanged on exit, but see Section 11.

IB – INTEGER.

On entry, IB must specify the first dimension of array B as declared in the calling (sub)program.
$IB \geq N$
Unchanged on exit.

N – INTEGER.

On entry, N must specify the order of matrix A.
Unchanged on exit.

M – INTEGER.

On entry, M must specify the number of lines of the matrix on either side of the diagonal.
Unchanged on exit.

IR – INTEGER.

On entry, IR must specify the number of right hand sides.
Unchanged on exit.

C – *real* array of DIMENSION (IC,p) where $p \geq IR$.

On successful exit, C will contain the IR solution vectors.

IC – INTEGER.

On entry, IC must specify the first dimension of array C as declared in the calling (sub)program.
$IC \geq N$
Unchanged on exit.

RL – *real* array of DIMENSION (IRL,p) where $p \geq M1$.

On successful exit, RL will contain the elements of the lower triangle of the Cholesky decomposition of A stored in the same form as A, except that the reciprocals of the diagonal elements are stored instead of the elements themselves.

IRL – INTEGER.

On entry, IRL must specify the first dimension of array RL as declared in the calling (sub)program.
$IRL \geq N$
Unchanged on exit.

M1 – INTEGER.

On entry, M1 must specify the value $(M+1)$.

Unchanged on exit.

IFAIL – INTEGER.

Before entry, IFAIL must be set to 0 or 1. For users not familiar with this parameter (described in Chapter P01) the recommended value is 0.

Unless the routine detects an error (see next section), IFAIL contains 0 on exit.

6.   Error Indicators and Warnings

Errors detected by the routine:–

IFAIL = 1

The band matrix is not positive definite, possibly due to rounding errors.

7.   Auxiliary Routines

Details are distributed to sites in machine-readable form.

8.   Timing

The time taken is approximately proportional to $N \times M^2$.

9.   Storage

There are no internally declared arrays.

10.   Accuracy

The accuracy of the computed solutions depend on the conditioning of the original matrix. For a detailed error analysis see [1], p. 54.

11.   Further Comments

This routine should only be used when $M \ll N$ since as M approaches N, it becomes less efficient to take advantage of the band form. If the routine is called with the same name for parameters B and C, then the solution vectors will overwrite the right hand sides.

12.   Keywords

Approximate Solution of Linear Equations,
Cholesky Decomposition,
Multiple Right Hand Sides,
Real Symmetric Positive Definite Band Matrix.

13.   Example

To solve the set of linear equations $AX = B$ where

$$A = \begin{pmatrix} 5 & -4 & 1 & & & & \\ -4 & 6 & -4 & 1 & & & \\ 1 & -4 & 6 & -4 & 1 & & \\ & 1 & -4 & 6 & -4 & 1 & \\ & & 1 & -4 & 6 & -4 & 1 \\ & & & 1 & -4 & 6 & -4 \\ & & & & 1 & -4 & 5 \end{pmatrix} \text{ and } B = \begin{pmatrix} 0 \\ 0 \\ 0 \\ 1 \\ 0 \\ 0 \\ 0 \end{pmatrix}$$

13.1.   Program Text

WARNING: This **single precision** example program may require amendment for certain implementations. The results produced may not be the same. If in doubt, please seek further advice (see **Essential Introduction** to the Library Manual).

```
C     F04ACF EXAMPLE PROGRAM TEXT
C     NAG COPYRIGHT 1975
C     MARK 4.5 REVISED
C
      REAL A(10,5), B(10,1), C(10,1), RL(10,5), TITLE(18)
      INTEGER NIN, NOUT, I, N, M, LR, M1, J, IA, IB, IC, IL, IFAIL
      DATA NIN /5/, NOUT /6/
      READ (NIN,99999) (TITLE(I),I=1,7)
      WRITE (NOUT,99997) (TITLE(I),I=1,6)
      N = 7
      M = 2
      LR = 1
      M1 = 3
      READ (NIN,99998) ((A(I,J),J=1,M1),B(I,1),I=1,N)
      IA = 10
      IB = 10
      IC = 10
      IL = 10
      IFAIL = 1
      CALL F04ACF(A, IA, B, IB, N, M, LR, C, IC, RL, IL, M1, IFAIL)
      IF (IFAIL.EQ.0) GO TO 20
      WRITE (NOUT,99996) IFAIL
      STOP
   20 WRITE (NOUT,99995) (C(I,1),I=1,N)
      STOP
99999 FORMAT (6A4, 1A3)
99998 FORMAT (4F5.0)
99997 FORMAT (4(1X/), 1H , 5A4, 1A3, 7HRESULTS/1X)
99996 FORMAT (25H0ERROR IN F04ACF IFAIL = , I2)
99995 FORMAT (10H0SOLUTIONS/(1H , F4.1))
      END
```

13.2.   Program Data

```
F04ACF EXAMPLE PROGRAM DATA
     0     0     5     0
     0    -4     6     0
     1    -4     6     0
     1    -4     6     1
     1    -4     6     0
     1    -4     6     0
     1    -4     5     0
```

13.3.   Program Results

```
F04ACF EXAMPLE PROGRAM RESULTS

SOLUTIONS
  4.0
  7.5
 10.0
 11.0
 10.0
  7.5
  4.0
```

(II) Commentary

Summary of information needed to run F04ACF

Purpose: This routine finds the unique solution of sets of linear simultaneous equations PC = B, where P is a symmetric positive-definite band matrix.

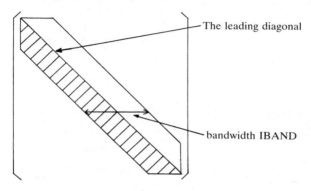

The leading diagonal

bandwidth IBAND

Comment:
A band matrix is one in which all the non-zero elements of a matrix lie in a band about the leading diagonal [see figure above].

The **routine name** with parameters is

F04ACF (A,IA,B,IB,N,M,IR,C,IC,RL,IRL,M1,IFAIL).

Housekeeping

Parameter	Meaning	Value needed before a routine call	To be printed after a routine call	Suggestions: size of array in declaration	Initial value
N [Integer]	The number of equations	√			
M [Integer]	(IBAND−1)/2, where IBAND is the bandwidth [see diagram]	√			M=(IBAND−1)/2
IR [Integer]	The number of right-hand sides	√			
A [Real array]	The elements of the matrix P described below	√		A(20,5)	
Ref. **§2.4** IA [Integer]	First dimension of A in declaration	√		IA=20	

continued

Table—continued

| | | | | Suggestions: | |
Parameter	Meaning	Value needed before a routine call	To be printed after a routine call	size of array in declaration	Initial value
B [Real array]	The right-hand sides B of PC=B	✓		B(20,4)	
IB [Integer]	First dimension of B in declaration	✓			IB=20
C [Real array]	The solution C in PC=B		✓	C(20,4)	
IC [Integer]	First dimension of C in declaration	✓			IC=20
Ref. §3.3 RL [Real array]	Workspace			RL(20,5)	
IRL [Integer]	First dimension of RL in declaration	✓			IRL=20
M1 (Integer)	(M + 1)	✓			M1=M+1
Ref. §3.2 IFAIL [Integer]	The usual error parameter	✓			IFAIL=0

Comments:

(i) The suggestions in the right-hand column of the table above would allow for a maximum of 20 equations, a maximum bandwidth of 9, with up to 4 right-hand sides. If you have more equations, a greater bandwidth, or more than 4 right-hand sides, then you will have to change this column, the subsequent program plan and the specimen program accordingly.

The following comments use the matrix

$$P = \begin{bmatrix} 9 & 2 & 0 & 0 & 0 \\ 2 & 3 & 2 & 0 & 0 \\ 0 & 2 & 5 & 1 & 0 \\ 0 & 0 & 1 & 7 & 2 \\ 0 & 0 & 0 & 2 & 10 \end{bmatrix}$$

as an illustration.

(ii) M must contain the number of elements on either side of the leading diagonal. In the case of the tri-diagonal matrix P above, there is one element on either side of the diagonal, so M should be given the value 1.

An alternative approach to finding M is just to count the total bandwidth IBAND (3 in the case above), and to get the program to find M using a statement

$$M = (IBAND-1)/2.$$

(iii) The reason for using the routine F04ACF rather than any other is that if you have a band symmetric positive definite matrix, then relatively little storage space is required. For instance, in the figure above, the only information which you have to supply is the numbers

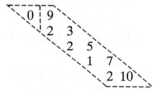

(Note that a zero has been added on the left in row 1 so that each row has the same number of elements.) This information is stored in an array A. So, in this case,

$$A(1,1) = 0.0, \quad A(1,2) = 9.0$$
$$A(2,1) = 2.0, \quad A(2,2) = 3.0$$
$$\vdots \qquad\qquad \vdots$$
$$A(5,1) = 2.0, \quad A(5,2) = 10.0$$

In general, A will have N rows and (M+1) columns.

(III) Program

Program plan

1. *Declare* REAL A(20,5), B(20,4), C(20,4), RL(20,5)
   INTEGER IA, IB, N, M, IR, IC, IRL, M1, IFAIL

2. *Read* N, IBAND (the total bandwidth), IR

3. *Set* M=(IBAND−1)/2; M1=M+1; IA=20; IB=20; IC=20;
   IRL=20; IFAIL=0

4. *Read* A, B

5. *Call* F04ACF

6. *Print* C

The following program is used to solve the equations given in the NAG routine document.

i.e.

$$\begin{bmatrix} 5 & -4 & 1 & & & & \\ -4 & 6 & -4 & 1 & & & \\ 1 & -4 & 6 & -4 & 1 & & \\ & 1 & -4 & 6 & -4 & 1 & \\ & & 1 & -4 & 6 & -4 & 1 \\ & & & 1 & -4 & 6 & -4 \\ & & & & 1 & -4 & 5 \end{bmatrix} \begin{bmatrix} x_1 \\ x_2 \\ x_3 \\ x_4 \\ x_5 \\ x_6 \\ x_7 \end{bmatrix} = \begin{bmatrix} 0 \\ 0 \\ 0 \\ 1 \\ 0 \\ 0 \\ 0 \end{bmatrix}$$

F04ACF specimen program

Ref.
§5.2

In the following program the data are read in from a data file.

```
C       FO4ACF: LINEAR SIMULTANEOUS EQUATIONS
C               (BAND SYMMETRIC POSITIVE-DEFINITE MATRIX)

        REAL A(20,5), B(20,4), C(20,4), RL(20,5)
        INTEGER IA, IB, N, M, IR, IC, IRL, Ml, IFAIL, IBAND,
     *          I, J

        OPEN (UNIT=20,FILE='F04ACF.DAT')

        IA = 20
        IB = 20
        IC = 20
        IRL = 20

        WRITE (6,*) 'THE NUMBER OF EQUATIONS'
        READ (20,*) N
        WRITE (6,*) N
        WRITE (6,*) 'THE TOTAL BANDWIDTH OF YOUR MATRIX'
        READ (20,*) IBAND
        WRITE (6,*) IBAND
        M = (IBAND-1)/2
        Ml = M + 1
        WRITE (6,*) 'THE LOWER HALF OF THE MATRIX'
        WRITE (6,*) '(CONSULT MANUAL FOR MORE DETAILS)'
        DO 10 I = 1, N
           READ (20,*) (A(I,J),J=1,M+1)
           WRITE (6,'(5F10.3)') (A(I,J),J=1,M+1)
 10     CONTINUE
        WRITE (6,*) 'THE NUMBER OF RIGHT HAND SIDES'
        READ (20,*) IR
        WRITE (6,*) IR
        WRITE (6,*) 'THE RIGHT HAND SIDES'
        DO 20 I = 1, N
           READ (20,*) (B(I,J),J=1,IR)
           WRITE (6,'(4F10.3)') (B(I,J),J=1,IR)
 20     CONTINUE

        IFAIL = 0
```

```
         CALL F04ACF(A,IA,B,IB,N,M,IR,C,IC,RL,IRL,M1,IFAIL)

         WRITE (6,*) 'THE SOLUTIONS ARE'
         DO 30 I = 1, N
           WRITE (6,'(4F10.3)') (C(I,J),J=1,IR)
    30   CONTINUE

         STOP

         END
```

F04ACF specimen run

```
THE NUMBER OF EQUATIONS
7
THE TOTAL BANDWIDTH OF YOUR MATRIX
5
THE LOWER HALF OF THE MATRIX
(CONSULT MANUAL FOR MORE DETAILS)
      0.000       0.000     5.000
      0.000      -4.000     6.000
      1.000      -4.000     6.000
      1.000      -4.000     6.000
      1.000      -4.000     6.000
      1.000      -4.000     6.000
      1.000      -4.000     5.000
THE NUMBER OF RIGHT HAND SIDES
1
THE RIGHT HAND SIDES
      0.000
      0.000
      0.000
      1.000
      0.000
      0.000
      0.000
THE SOLUTIONS ARE
      4.000
      7.500
     10.000
     11.000
     10.000
      7.500
      4.000
```

10.7 Least-squares minimization: E04FDF

(I)   **E04FDF – NAG FORTRAN Library Routine Document**

NOTE: before using this routine, please read the appropriate implementation document to check the interpretation of **bold italicised** terms and other implementation–dependent details. The routine name may be precision–dependent.

1.   **Purpose**

E04FDF is an easy-to-use algorithm for finding an unconstrained minimum of a sum of squares of M nonlinear functions in N variables ($M \geq N$). No derivatives are required.

It is intended for functions which are continuous and which have continuous first and second derivatives (although it will usually work even if the derivatives have occasional discontinuities).

2.   **Specification**

```
      SUBROUTINE E04FDF(M,N,X,FSUMSQ,IW,LIW,W,LW,IFAIL)
C     INTEGER    M,N,IW(LIW),LIW,LW,IFAIL
C     real       X(N),FSUMSQ,W(LW)
```

3.   **Description**

This routine is essentially identical to the subroutine LSNDN1 in the National Physical Laboratory Algorithms Library. It is applicable to problems of the form

$$\text{Minimize } F(X) = \sum_{i=1}^{M} [f_i(X)]^2$$

where $X = (X_1, X_2,...,X_N)^T$ and $M \geq N$. (The functions $f_i(X)$ are often referred to as 'residuals'.) The user must supply a subroutine LSFUN1 to evaluate functions $f_i(X)$ at any point X.

From a starting point supplied by the user, a sequence of points is generated which is intended to converge to a local minimum of of the sum of squares. These points are generated using estimates of the curvature of F(X).

4.   **References**

[1]   GILL, P.E. and MURRAY, W.
       Algorithms for the solution of non-linear least squares problem.
       SIAM Journal on Numerical Analysis, 15, pp. 977–992, 1978.

5.   **Parameters**

5.1.   **Parameters**

M – INTEGER.

N – INTEGER.

On entry, M must specify the number of residuals, $f_i(X)$, and N, the number of variables, X_j.

$1 \leq N \leq M$.

M and N are unchanged on exit.

X – **real** array of DIMENSION at least (N).

Before entry, X(j) must be set by the user to a guess at the j(th) component of the position of the minimum (j = 1,2,...,N).

On exit, X contains the lowest point found during the calculations. Thus, if $IFAIL = 0$ on exit, $X(j)$ is the j(th) component of the position of the minimum.

FSUMSQ – *real*.

On exit, FSUMSQ contains the value of the sum of squares, $F(X)$, corresponding to the final point stored in X.

IW – INTEGER array of DIMENSION at least (1).

Used as workspace.

LIW – INTEGER.

On entry, LIW must specify the acutal length of IW as declared in the calling (sub)program.

$LIW \geq 1$

Unchanged on exit.

W – *real* array of DIMENSION (LW).

Used as workspace.

LW – INTEGER.

On entry, LW must specify the actual length of W as declared in the calling (sub)program.

$LW \geq 7 \times N + N \times N + 2 \times M \times N + 3 \times M + N \times (N-1)/2$ if $N > 1$,
$LW \geq 9 + 5 \times M$ if $N = 1$.
Unchanged on exit.

IFAIL – INTEGER.

Before entry, IFAIL must be set to 0 or 1. Users who are unfamiliar with this parameter should refer to Chapter P01 for details. Unless the routine detects an error (see Section 6), IFAIL contains 0 on exit.

For this routine, because the values of output parameters may be useful even if IFAIL $\neq$ 0 on exit, users are recommended to set IFAIL to 1 before entry. **It is then essential to test the value of IFAIL on exit.**

5.2.   User-supplied Routines

LSFUN1 – SUBROUTINE.

This routine must be supplied by the user to calculate the vector of values $f_i(X)$ at any point X. Since the routine is not a parameter to E04FDF, it **must** be called LSFUN1. It should be tested separately before being used in conjunction with E04FDF (see Chapter Introduction).

The specification is:

```
SUBROUTINE LSFUN1(M,N,XC,FVECC)
INTEGER    M,N
real       XC(N),FVECC(M)
```

M – INTEGER.
N – INTEGER.

On entry, M and N contain the number of residuals and variables, respectively. Their values must not be changed in LSFUN1.

XC – *real* array of DIMENSION (N)

> On entry, XC contains the point at which the values of the f(i) are required. LSFUN1 must not change the values in XC.

FVECC – *real* array of DIMENSION (M).

> On exit, FVECC(i) must contain the value of f(i) at the point XC, for i = 1,2,...,M.

6.   Error Indicators and Warnings

Errors or warnings specified by the routine:–

IFAIL = 1

> On entry,   N < 1,
> or   M < N,
> or   LIW < 1,
> or   LW < 7×N + N×N + 2×M×N + 2×M + N×(N−1)/2 when N > 1,
> or   LW < 9 + 5×M when N = 1,
> or   IFAIL < 0 or IFAIL > 1.

IFAIL = 2

> There have been 400 × N calls of LSFUN1, yet the algorithm does not seem to have converged. This may be due to an awkward function or to a poor starting point, so it is worth restarting E04FDF from the final point held in X.

IFAIL = 3

> The final point does not satisfy the conditions for acceptance as a minimum, but no lower point could be found.

IFAIL = 4

> An auxiliary routine has been unable to complete a singular value decomposition in a reasonable number of sub-iterations.

IFAIL = 5
IFAIL = 6
IFAIL = 7
IFAIL = 8

> There is some doubt about whether the point X found by E04FDF is a minimum of F(X). The degree of confidence in the result decreases as IFAIL increases. Thus when IFAIL = 5, it is probable that the final X gives a good estimate of the position of a minimum, but when IFAIL = 8 it is very unlikely that the routine has found a minimum.

If the user is not satisfied with the result (e.g. because IFAIL lies between 3 and 8), it is worth restarting the calculations from a different starting point (not the point at which the failure occurred) in order to avoid the region which caused the failure. Repeated failure may indicate some defect in the formulation of the problem.

7.   Auxiliary Routines

Details are distributed to sites in machine-readable form.

8.   Timing

The number of iterations required depends on the number of variables, the number of residuals and their behaviour, and the distance of the starting point from the solution. The number of multiplications performed per iteration of E04FDF varies, but for $M \gg N$ is approximately $N \times M^2 + O(N^3)$. In addition, each iteration makes at least $N + 1$ calls of LSFUN1. So, unless the residuals can be evaluated very quickly, the run time will be dominated by the time spent in LSFUN1.

9.   Storage

There are no internally declared arrays.

10.   Accuracy

If the problem is reasonably well scaled and a successful exit is made, then, for a computer with a wordlength of t decimals, one would expect to get about $t/2 - 1$ decimals accuracy in the components of X and between $t - 1$ (if $F(X)$ is of order 1 at the minimum) and $2t - 2$ (if $F(X)$ is close to zero at the minimum) decimals accuracy in $F(X)$.

11.   Further Comments

Ideally, the problem should be scaled so that the minimum value of the sum of squares is in the range $(0, +1)$, and so that at points a unit distance away from the solution the sum of squares is approximately a unit value greater than at the minimum. It is unlikely that the user will be able to follow these recommendations very closely, but it is worth trying (by guesswork), as sensible scaling will reduce the difficulty of the minimization problem, so that E04FDF will take less computer time.

12.   Keywords

Corrected Gauss–Newton,
Easy-to–Use,
Finite Differences,
Function-only Method,
Nonlinear Least Squares,
Unconstrained.

13.   Example

To find least squares estimates of X_1, X_2 and X_3 in the model

$$Y = X_1 + \frac{T_1}{X_2 T_2 + X_3 T_3}$$

using the 15 sets of data given in the following table.

Y	T_1	T_2	T_3
0.14	1.0	15.0	1.0
0.18	2.0	14.0	2.0
0.22	3.0	13.0	3.0
0.25	4.0	12.0	4.0
0.29	5.0	11.0	5.0
0.32	6.0	10.0	6.0
0.35	7.0	9.0	7.0
0.39	8.0	8.0	8.0
0.37	9.0	7.0	7.0
0.58	10.0	6.0	6.0
0.73	11.0	5.0	5.0
0.96	12.0	4.0	4.0
1.34	13.0	3.0	3.0
2.10	14.0	2.0	2.0
4.39	15.0	1.0	1.0

The program uses (0.5, 1.0, 1.5) as the initial guess at the position of the minimum.

13.1. Program Text

WARNING: This **single precision** example program may require amendment for certain implementations. The results produced may not be the same. If in doubt, please seek further advice (see **Essential Introduction** to the Library Manual).

```
C       E04FDF EXAMPLE PROGRAM TEXT.
C       MARK 7 RELEASE. NAG COPYRIGHT 1978.
C       .. ARRAYS IN COMMON ..
        REAL T(15,3), Y(15)
C       ..
C       .. LOCAL SCALARS ..
        REAL FSUMSQ
        INTEGER I, IFAIL, J, LIW, LW, M, N, NIN, NOUT
C       .. LOCAL ARRAYS ..
        REAL TITLE(7), W(168), X(3)
        INTEGER IW(1)
C       .. SUBROUTINE REFERENCES ..
C       E04FDF
C       ..
        COMMON Y, T
        DATA NIN, NOUT /5,6/
        READ (NIN,99994) TITLE
        WRITE (NOUT,99995) (TITLE(I),I=1,6)
        M = 15
        N = 3
C       OBSERVATIONS OF TJ (J = 1, 2, 3) ARE HELD IN T(I, J)
C       (I = 1, 2, . . . , 15)
        DO 20 I=1,M
            READ (NIN,99999) Y(I), (T(I,J),J=1,N)
     20 CONTINUE
        X(1) = 0.5
        X(2) = 1.0
        X(3) = 1.5
        LIW = 1
C       7*N + N*N + 2*M*N + 3*M + N*(N - 1)/2 IS 168
        LW = 168
        IFAIL = 1
        CALL E04FDF(M, N, X, FSUMSQ, IW, LIW, W, LW, IFAIL)
C       SINCE IFAIL WAS SET TO 1 BEFORE ENTERING E04FDF, IT IS
C       ESSENTIAL TO TEST WHETHER IFAIL IS NON-ZERO ON EXIT
        IF (IFAIL.NE.0) WRITE (NOUT,99998) IFAIL
```

```
       IF (IFAIL.EQ.1) GO TO 40
       WRITE (NOUT,99997) FSUMSQ
       WRITE (NOUT,99996) (X(J),J=1,N)
    40 STOP
 99999 FORMAT (F5.2, 3F5.1)
 99998 FORMAT (///16H ERROR EXIT TYPE, I3, 22H - SEE ROUTINE DOCUMEN,
      * 1HT)
 99997 FORMAT (///31H ON EXIT, THE SUM OF SQUARES IS, F12.4)
 99996 FORMAT (13H AT THE POINT, 3F12.4)
 99995 FORMAT (4(1X/), 1H , 5A4, 1A3, 7HRESULTS/1X)
 99994 FORMAT (6A4, 1A3)
       END
C
       SUBROUTINE LSFUN1(M, N, XC, FVECC)
C      ROUTINE TO EVALUATE THE RESIDUALS
C      .. SCALAR ARGUMENTS ..
       INTEGER M, N
C      .. ARRAY ARGUMENTS ..
       REAL FVECC(M), XC(N)
C      ..
C      .. ARRAYS IN COMMON ..
       REAL T(15,3), Y(15)
C      ..
C      .. LOCAL SCALARS ..
       INTEGER I
C      ..
       COMMON Y, T
       DO 20 I=1,M
          FVECC(I) = XC(1) + T(I,1)/(XC(2)*T(I,2)+XC(3)*T(I,3)) -
      *   Y(I)
    20 CONTINUE
       RETURN
       END
```

13.2. Program Data

```
E04FDF EXAMPLE PROGRAM DATA
 0.14  1.0 15.0  1.0
 0.18  2.0 14.0  2.0
 0.22  3.0 13.0  3.0
 0.25  4.0 12.0  4.0
 0.29  5.0 11.0  5.0
 0.32  6.0 10.0  6.0
 0.35  7.0  9.0  7.0
 0.39  8.0  8.0  8.0
 0.37  9.0  7.0  7.0
 0.58 10.0  6.0  6.0
 0.73 11.0  5.0  5.0
 0.96 12.0  4.0  4.0
 1.34 13.0  3.0  3.0
 2.10 14.0  2.0  2.0
 4.39 15.0  1.0  1.0
```

13.3. Program Results

```
E04FDF EXAMPLE PROGRAM RESULTS

ON EXIT, THE SUM OF SQUARES IS     0.0082
AT THE POINT      0.0824      1.1330      2.3437
```

(II) Commentary

Summary of information needed to run E04FDF

Purpose: The routine finds the *x*-values which minimize the sum

$$S = f_1^2 + f_2^2 + \ldots + f_m^2$$

where $f_1, f_2, \ldots, f_m$ are functions of the variables $x_1, x_2, \ldots, x_n$. This problem has the restriction that $m \geq n$, where m is the number of functions, and n is the number of *x*-variables.

Comments:

(a) A straightforward application of this routine would be to find the values of x_1 and x_2 which minimize

$$S = (x_1^2 + 2x_2)^2 + \cos^2 x_2 + (x_2^2 - 3x_1 x_2)^2 .$$

Here, $f_1(x_1,x_2) = x_1^2 + 2x_2$

$\quad\quad\ f_2(x_1,x_2) = \cos x_2$

and $\quad f_3(x_1,x_2) = x_2^2 + 3x_1 x_2 .$

However, this is not the most common application. The routine is normally used to find how well a given set of data fits some non-linear equation of your choice.

Suppose, for instance that you have a table of 15 sets of readings

y	0.14	0.18	0.22	0.25	0.29	0.32	0.35	0.39	0.37	0.58	0.73	0.96	1.34	2.10	4.39
u	1.0	2.0	3.0	4.0	5.0	6.0	7.0	8.0	9.0	10.0	11.0	12.0	13.0	14.0	15.0
v	15.0	14.0	13.0	12.0	11.0	10.0	9.0	8.0	7.0	6.0	5.0	4.0	3.0	2.0	1.0
w	1.0	2.0	3.0	4.0	5.0	6.0	7.0	8.0	7.0	6.0	5.0	4.0	3.0	2.0	1.0

where u, v, and w are independent variables. Suppose further that you suspect that there is a relationship of the form

$$y \simeq a + \frac{u}{bv + cw} .$$

where a, b, and c are unknown constants. Then, using the method of least-squares, you can form the residuals

$$f_i = y_i - \left[a + \frac{u_i}{bv_i + cw_i} \right] \quad i = 1, \ldots 15 \quad\quad\quad (10.1)$$

and find the values of a, b, and c which minimize

$$S = f_1^2 + f_2^2 + \ldots + f_{15}^2 . \tag{10.2}$$

The minimum value of S will tell you how good the fit is.

To use E04FDF to do this problem, you must set up a function corresponding to each of the m readings or observations.

The problem, described opposite, is the specimen problem given in the NAG routine document. There are two minor differences. The unknown constants a, b, and c in the problem above are called $X(1)$, $X(2)$ and $X(3)$ in the NAG routine document. Use of an array X allows you to have as many unknown constants as your problem requires. Also, the i(th) readings u_i, v_i and w_i in the problem above are called $T(I,1)$, $T(I,2)$ and $T(I,3)$ in the NAG routine document. Again, use of an array T allows for generalization.

It is not necessary that the number of independent variables u, v, w should be the same as the number of unknown constants a, b, c, although this happens to be so in the problem above.

(b) The restriction $m \geqslant n$ has the straightforward interpretation that there must be as many, or more functions than x-variables. In terms of the problem above, this means that there must be as many, or more, sets of readings than unknown constants.

The **routine name** with parameters is

E04FDF (M, N, X, FSUMSQ, IW, LIW, W, LW, IFAIL).

Housekeeping

(i) *for E04FDF*:

Parameter	Meaning	Value needed before a routine call	To be printed after a routine call	Suggestions: size of array in declaration	Initial value
M [Integer]	The number of functions (or sets of readings)	✓			
N [Integer]	The number of variables (or unknown constants)	✓			

continued

Table – *continued*

	Parameter	Meaning	Value needed before a routine call	To be printed after a routine call	Suggestions: size of array in declaration	Initial value
	X [Real array]	*Before* a routine call: a guess at the x-values *After* a routine call: the x-values which minimize S	✓	✓	X(6)	
	FSUMSQ [Real]	The minimum value of S		✓		
	IW [Integer array]	Arrays used as workspace			IW(1)	
	W [Real array]				W(1600)	
Ref. **§2.4**	LIW LW [Integer]	The lengths of IW and W in the declaration	✓ ✓			LIW=1 LW=1600
Ref. **§3.2**	IFAIL [Integer]	The usual error parameter	✓			IFAIL=1

Comment:
It is suggested that you set IFAIL = 1 initially when you run this program. The reason for this is that the computed x-values might be of interest, even if IFAIL is not zero after a routine call. Note in the specimen program that IFAIL has been printed, but not tested, so be careful to take note of its value.

(ii) *for LSFUN1*:

The **purpose** of this routine is to define the functions $f_1, f_2, \ldots, f_m$.

The **routine name** with parameters is

LSFUN1(M, N, XC, FVECC).

Comments:
(i) The routine LSFUN1 is called by E04FDF. Note that LSFUN1 is not a parameter of E04FDF, and so must be given the name LSFUN1.
(ii) Take heed of the warning in the NAG routine document not to assign values to M, N, or XC in the subroutine LSFUN1. The routine E04FDF supplies these values when required.

Parameter	Meaning	Suggested size of arrays in declaration
M [Integer]	Number of functions (or sets of readings)	
N [Integer]	Number of variables (or unknown constants)	
XC [Real array]	Values of $x_1, x_2, \ldots,$ x_n. These are assigned by E04FDF.	XC(N)
FVECC [Real array]	Array where the functions $f_1, f_2,$ $\ldots, f_m$ are defined	FVECC(M)

Ref.
§2.4

Comment:

As M and N are parameters of the subroutine LSFUN1, you can use the declaration

REAL XC(N), FVECC(M)

in LSFUN1.

A subroutine which defines the functions

$$f_i = y_i - \left(x_1 + \frac{t_{i1}}{x_2 t_{i2} + x_3 t_{i3}}\right) \quad i = 1, 2, \ldots, 15$$

needed for the problem given in (10.1) and (10.2) could be written as follows:

```
      SUBROUTINE LSFUN1(M, N, XC, FVECC)
      COMMON T(100,5), Y(100)
      INTEGER M, N, I
      REAL XC(N), FVECC(M)
      DO 10 I = 1, M
         FVECC(I) = Y(I) - XC(1)
     *              - T(I,1)/(XC(2)*T(I,2)+XC(3)*T(I,3))
   10 CONTINUE
      RETURN
      END
```

Ref.
§2.5

Comment:

Note the need for the COMMON statement in the subroutine above. Values for both Y and T are needed in LSFUN1 to define the functions. As neither Y or T are parameters of LSFUN1, both the calling program and LSFUN1 need a declaration

COMMON T(,), Y( ) .

Moreover, it is essential that the dimensions of the arrays T and Y should be the same in both the calling program and LSFUN1.

Finally, the suggestions given in the right-hand columns of the two tables above, along with COMMON statements

COMMON T(100,5), Y(100)

would allow for
(a) up to 100 functions (or sets of readings)
(b) up to 5 t-variables in the calling program and in LSFUN1 and
(c) up to 6 x-variables (or unknown constants).

If your program has more functions, or more x- or t-variables than this, then you will have to change the suggestions, the program plan and the specimen program accordingly.

(III) Program

Program plan

1. *Declare*	COMMON T(100,5), Y(100)
	REAL X(6), FSUMSQ, W(1600)
	INTEGER N, M, IW(1), LIW, LW, IFAIL
2. *Read*	M, N, X
3. *Read*	K (the number of t-variables), Y, T
4. *Set*	LW=1600; LIW=1; IFAIL=1
5. *Call*	E04FDF
6. *Print*	IFAIL, X, FSUMSQ
7. *Write a subroutine* LSFUN	

The following program is used to find X(1), X(2) and X(3) for the problem given in the NAG routine document.

E04FDF specimen program

Ref.
§5.2 In the following program, the data are read in from a data file.

```
C       E04FDF: MINIMIZATION OF A FUNCTION
C               (LEAST SQUARES PROBLEM)

        REAL X(6), FSUMSQ, W(1600), T(100,5), Y(100)
        INTEGER M, N, IW(1), LIW, LW, IFAIL, I, J, K
        COMMON T, Y
```

```
      OPEN (UNIT=20,FILE='E04FDF.DAT')

      LW = 1600
      LIW = 1

      WRITE (6,*) 'THE NUMBER OF OBSERVATIONS'
      READ (20,*) M
      WRITE (6,*) M
      WRITE (6,*)
    * 'THE NUMBER OF T VALUES IN EACH OBSERVATION'
      READ (20,*) K
      WRITE (6,*) K
      WRITE (6,*) 'THE DATA: ONE OBSERVATION PER LINE'
      WRITE (6,*) 'Y VALUE FOLLOWED BY ', K, ' T VALUES'
      DO 10 I = 1, M
         READ (20,*) Y(I), (T(I,J),J=1,K)
         WRITE (6,'(1X,6E12.3)') Y(I), (T(I,J),J=1,K)
 10   CONTINUE
      WRITE (6,*) 'ENTER THE NUMBER OF UNKNOWN COEFFICIENTS'
      READ (5,*) N
      WRITE (6,*)
    * 'ENTER A GUESS AT THE MINIMIZING COEFFICENTS'
      READ (5,*) (X(I),I=1,N)

      IFAIL = 1

      CALL E04FDF(M,N,X,FSUMSQ,IW,LIW,W,LW,IFAIL)

      WRITE (6,*) 'IFAIL=', IFAIL
      WRITE (6,*) 'THE UNKNOWN COEFFICIENTS'
      DO 20 I = 1, N
         WRITE (6,*) 'X(', I, ') = ', X(I)
 20   CONTINUE
      WRITE (6,*) 'WITH RESIDUAL SUM OF SQUARES ', FSUMSQ

      STOP

      END

      SUBROUTINE LSFUN1(M,N,XC,FVECC)
      INTEGER M, N, I
      REAL XC(N), FVECC(M), T(100,5), Y(100)
      COMMON T, Y
      DO 10 I = 1, M
         FVECC(I) = Y(I) - XC(1)
    *              - T(I,1)/(XC(2)*T(I,2)+XC(3)*T(I,3))
 10   CONTINUE
      RETURN
      END
```

continued

E04FDF specimen run

```
THE NUMBER OF OBSERVATIONS
15
THE NUMBER OF T VALUES IN EACH OBSERVATION
3
THE DATA: ONE OBSERVATION PER LINE
Y VALUE FOLLOWED BY 3 T VALUES
        0.140E+00    0.100E+01    0.150E+02    0.100E+01
        0.180E+00    0.200E+01    0.140E+02    0.200E+01
        0.220E+00    0.300E+01    0.130E+02    0.300E+01
        0.250E+00    0.400E+01    0.120E+02    0.400E+01
        0.290E+00    0.500E+01    0.110E+02    0.500E+01
        0.320E+00    0.600E+01    0.100E+02    0.600E+01
        0.350E+00    0.700E+01    0.900E+01    0.700E+01
        0.390E+00    0.800E+01    0.800E+01    0.800E+01
        0.370E+00    0.900E+01    0.700E+01    0.700E+01
        0.580E+00    0.100E+02    0.600E+01    0.600E+01
        0.730E+00    0.110E+02    0.500E+01    0.500E+01
        0.960E+00    0.120E+02    0.400E+01    0.400E+01
        0.134E+01    0.130E+02    0.300E+01    0.300E+01
        0.210E+01    0.140E+02    0.200E+01    0.200E+01
        0.439E+01    0.150E+02    0.100E+01    0.100E+01
ENTER THE NUMBER OF UNKNOWN COEFFICIENTS
3
ENTER A GUESS AT THE MINIMIZING COEFFICENTS
0.5  1.0  1.5
IFAIL=0
THE UNKNOWN COEFFICIENTS
X(1) = 8.2411979E-02
X(2) = 1.133078
X(3) = 2.343655
WITH RESIDUAL SUM OF SQUARES 8.2148778E-03
```

Index